T0202028

Supramolecular Chemistry

Supramolecular Chemistry: Fundamentals and Applications

SECOND EDITION

Paul D. Beer
University of Oxford U.K.

Timothy A. Barendt
University of Birmingham U.K.

Jason Y. C. Lim
Institute of Materials Research and Engineering, Singapore

OXFORD
UNIVERSITY PRESS

OXFORD
UNIVERSITY PRESS

Great Clarendon Street, Oxford, OX2 6DP,
United Kingdom

Oxford University Press is a department of the University of Oxford.
It furthers the University's objective of excellence in research, scholarship,
and education by publishing worldwide. Oxford is a registered trade mark of
Oxford University Press in the UK and in certain other countries

© Paul Beer, Timothy Barendt and Jason Lim 2022

The moral rights of the authors have been asserted

Impression: 1
All rights reserved. No part of this publication may be reproduced, stored in
a retrieval system, or transmitted, in any form or by any means, without the
prior permission in writing of Oxford University Press, or as expressly permitted
by law, by licence or under terms agreed with the appropriate reprographics
rights organization. Enquiries concerning reproduction outside the scope of the
above should be sent to the Rights Department, Oxford University Press, at the
address above

You must not circulate this work in any other form
and you must impose this same condition on any acquirer

Published in the United States of America by Oxford University Press
198 Madison Avenue, New York, NY 10016, United States of America

British Library Cataloguing in Publication Data
Data available

Library of Congress Control Number: 2021937962

ISBN 978-0-19-883284-3

Printed in Great Britain by
Bell & Bain Ltd., Glasgow

Links to third party websites are provided by Oxford in good faith and
for information only. Oxford disclaims any responsibility for the materials
contained in any third party website referenced in this work.

Preface

This primer serves as an introductory text to supramolecular chemistry, focusing on the fundamental principles and key applications of this exciting interdisciplinary area of science. These are presented through a wide range of examples of supramolecular complexes, assemblies and architectures, with classical examples complemented by case studies from the very cutting-edge of supramolecular chemistry research.

Many will be familiar with the 1999 primer *Supramolecular Chemistry* authored by Paul Beer, Philip Gale and David Smith. *Supramolecular Chemistry: Fundamentals and Applications* takes a new approach to the subject, designed to be used as a basis for a contemporary teaching course for final year undergraduate and graduate students. The primer consists of chapters on the binding of charged and neutral guest molecules, supramolecular self-assembly and, in light of the 2016 Nobel Prize in Chemistry, a chapter on mechanically interlocked molecules. All the chapters are fully illustrated to help visualise the relatively complex molecular structures. A glossary is also provided for key terminology, as well as links to further resources on each subject. To encourage active learning, each chapter concludes with questions for students to test critical knowledge and practice problem solving.

The authors would like to thank the following people for their very helpful comments and suggestions with various sections of this book: Prof. David Leigh (Manchester, UK), Dr Nicholas White (ANU, Australia), Dr. Marion Kieffer (Switzerland).

Paul D. Beer, Timothy A. Barendt and Jason Y. C. Lim
Oxford, Birmingham and Singapore
April 2021

Contents

Introduction to supramolecular chemistry

1.1 Introduction

Throughout the twentieth century, great advances in synthetic chemistry have provided ingenious means of constructing covalent bonds between atoms to form ever-more complex molecules. Yet, it is easy to overlook the fact that discrete molecules also do interact *non-covalently* with each other via intermolecular forces. In fact, in some cases, they can assemble spontaneously to form super-molecules whose intricacies and complexities rival or even exceed any purely covalently bonded structure! The field of supramolecular chemistry, born of this realization, has been defined by 1987 Nobel laureate Jean-Marie Lehn as 'the chemistry of the intermolecular bond, covering the structures and functions of the entities formed by association of two or more chemical species'. Supramolecular chemistry broadly comprises of two main themes: (1) host–guest chemistry, the selective recognition and binding of one chemical entity to another; and (2) self-assembly, the spontaneous association of small molecules into larger and more complex structures (Figure 1.1). Indeed, the control of intermolecular interactions codified through concepts of supramolecular chemistry virtually underpins every aspect of modern chemistry today, and has been central to modern developments in molecular devices, sensors, non-covalent polymers, drug discovery and pharmaceutics, the consumer-care industry, soft materials, and catalysis, to name a few.

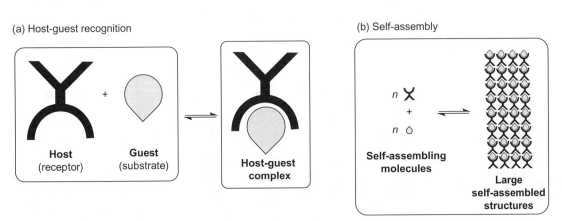

Figure 1.1 Schematic illustration describing the concepts of (A) host–guest recognition and (B) self-assembly.

Given its numerous applications, supramolecular chemistry is inherently a multi-disciplinary field, cutting across the traditional sub-divisions of organic, inorganic, biological, and physical chemistry. While organic and inorganic chemistry are required for the synthesis of the pre-designed supramolecular building blocks or 'synthons', physical chemistry techniques need to be employed to study their properties (nature of interaction, binding strength, molecular responses). Biological chemistry can provide the inspiration for the design of supramolecular molecules from nature, and translate discoveries into biological applications. In addition to these, computational chemists can offer insights regarding the interactions on the molecular scale, and modern machine-learning algorithms provide increasingly powerful predictive capabilities on design of supramolecular systems and their macroscopic behaviour. We hope that the examples and concepts described in the remaining pages of this primer will stimulate the imagination and creativity of budding supramolecular chemists, driving future innovations in diverse scientific endeavours informed by supramolecular concepts.

1.2 Supramolecular chemistry and molecular recognition in nature

Nature provides countless examples par excellence of functional supramolecular chemistry, which is essential for driving biomolecule self-assembly (proteins, DNA) and enzyme-substrate interactions. The self-assembly of antiparallel strands of DNA into double helices is driven by the highly specific *complementary* hydrogen bonding between pyrimidine (cytosine, thymine) and purine bases (guanine, adenine) (Figure 1.2A). Adenine (A) and thymine (T) form two hydrogen bonds with each other, whilst guanine (G) and cytosine (C) are held by three. Due to the spatial and geometric

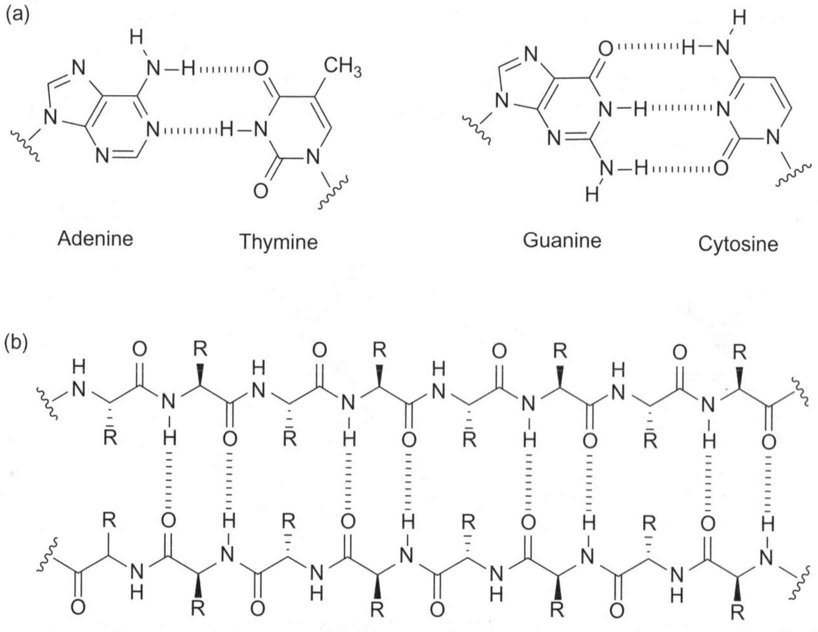

(a)

Adenine   Thymine   Guanine   Cytosine

(b)

Figure 1.2 Complementary hydrogen bonding (dashed lines) between (A) A–T and C–G bases on DNA and (B) amino acid residues in anti-parallel β-pleated sheets.

arrangements of the hydrogen bond donor and acceptor groups on each base, A–T and C–G base pairs form almost exclusively and mismatches rarely occur. This complementary base-pairing, which extends throughout the entire structure of DNA, ensures that one strand is complementary to the other, and is ultimately responsible for DNA replication. Other than DNA, complementary hydrogen bonding is also responsible for the folding of polypeptide chains into important secondary structures such as β-pleated sheets (Figure 1.2B).

These examples illustrate the fundamental supramolecular concept of **complementarity**, which maximizes the strength of non-covalent forces between interacting species. In enzymes, substrate specificity is achieved as their active sites are exquisitely complementary in shape, size, and geometry to the substrate (or the transition state of the catalysed reaction). This idea was described by Emil Fischer in 1894 as the 'lock and key' principle, where the 'lock' and 'key' represent the enzyme and substrate respectively, and only the right 'key' can fit into the 'lock'. However, it is also important that the enzyme's active site is held in a conformation that already pre-disposes it towards substrate binding, such that minimal energetically demanding conformational changes need to take place. This concept, termed **preorganization**, is essential to the design of many supramolecular systems described in the following pages.

The phosphate-binding protein of the *Halomonadaceae* bacteria strain GFAJ-1 is an excellent example of how complementarity and preorganization act together to bring about substrate selectivity. The bacterium, which thrives in the arsenate-rich waters of Mono Lake, USA, is able to discriminate HPO_4^{2-} with a 4500-fold selectivity over $HAsO_4^{2-}$ in water. This is especially impressive considering that the two anions share the same tetrahedral geometry, have identical charge, and almost identical pK_a values and hydration enthalpies. The origin of selectivity, which was only elucidated in 2012, arises from a subtle 4% size difference between these two anions. Due to the highly preorganized binding site, held in place by the overall protein structure, this anion size difference results in a steric clash by $HAsO_4^{2-}$ with a leucine residue (Leu9) in the binding site, which distorts a hydrogen bond from the –OH group of the anion to a nearby aspartate (Asp62) side chain (Figure 1.3A). In contrast, all interactions are

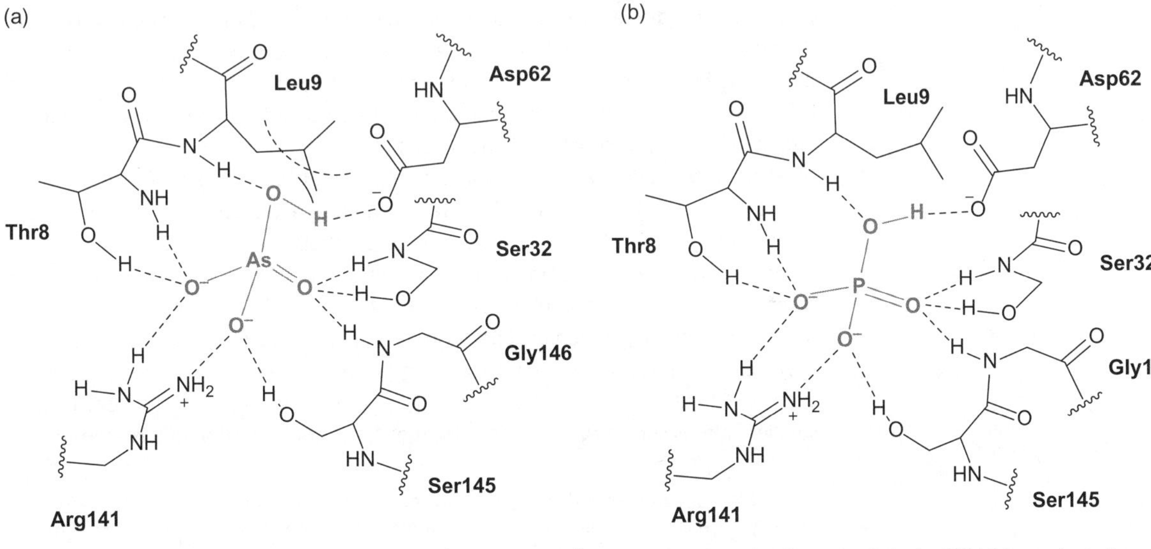

Figure 1.3 Origin of selectivity between (A) $HAsO_4^{2-}$ and (B) HPO_4^{2-} by the phosphate binding protein in the GFAJ-1 bacteria strain (M. Elias *et. al.*, *Nature*, 2012, **491**, 134).

perfectly orientated and complementary for HPO_4^{2-}, resulting in its preferred binding (Figure 1.3B). The synergy of complementarity and preorganization also accounts for the arguably even more subtle ability of enzymes to distinguish between chiral substrates and bind one enantiomer/diastereoisomer selectively over others. Given that these stereoisomers are identical in all aspects other than their 3D orientation of substituents around a chiral centre, selectivities in excess of 100-fold are all the more remarkable.

1.3 Design principles

The chelate, macrocyclic, and cryptate effects

To design successful supramolecular synthons which are able to bind strongly to each other or to guest molecules, supramolecular chemists rely on a set of design principles. With origins in transition metal coordination chemistry, the **chelate effect** is often exploited, where the guest binding is stabilized by the presence of multiple binding sites on the host. Consider the binding constants[1] for the complexes of $[Cu(H_2O)_6]^{2+}$ with the ligands NH_3, ethylene diamine (en), 1,4,7,10-triethylene tetraamine (trien), 1,4,8,11-triethylene tetraamine (2,3,2-tet) and cyclam (Figure 1.4A), where a greater number of chelating nitrogen atoms on the ligand result in notably larger binding constant values. This effect arises from various factors. Enthalpically, having the polar amino groups covalently tethered together overcomes part of their mutual repulsion experienced for separate NH_3 ligands, energetically favouring coordination. For the multidentate ligands, the coordination of one amino group to Cu^{2+} also brings the other amino groups on the ligand into closer proximity with the metal centre and facilitates binding. Entropically, multidentate ligand binding results in an overall greater number of displaced individual water molecules compared with NH_3. For NH_3, four ammonia ligands displace four inner-sphere water molecules, giving no net change in the number of free species in solution. During the coordination of the bidentate en ligand, two ligand molecules displace four water molecules to give a net increase of two free species. Coordination of each of the other tetradentate ligands results in four additional displaced free water molecules. Hence, binding of ligands with greater denticities increases the entropy of the system to a larger extent, lowering $\Delta G°$ and increasing binding constant magnitudes. More subtly, the covalent bridges between the amino groups also increase their basicities compared with NH_3 due to inductive effects, further increasing their donor capabilities.

The astute reader will have noticed that 2,3,2-tet forms more stable Cu^{2+} complexes than trien, despite both being tetradentate ligands. This has been attributed to the reduced ring strain in the six-membered chelate rings of 2,3,2-tet with Cu^{2+} compared with the five-membered rings of trien. It should be noted, however, that ring strains in transition metal complexes vary with the size of the metal cation. While four-membered rings (e.g. bidentate chelation by both oxygen atoms of acetate) are invariably highly strained, a five-membered ring (as in en) is preferable for larger cations such as Pb^{2+}, and six- or seven-membered rings are more favourable for coordinating smaller cations such as Cu^{2+}. For small cations, the conformation of a six-membered

---

[1] The term 'binding constant' or (K_a) can be used interchangeably with 'association constant', 'stability constant' or 'formation constant' to indicate the strength of host-guest interactions.

(a)

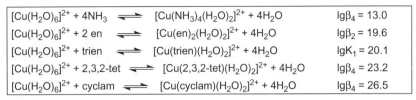

$[Cu(H_2O)_6]^{2+} + 4NH_3$ ⇌	$[Cu(NH_3)_4(H_2O)_2]^{2+} + 4H_2O$	$lg\beta_4 = 13.0$
$[Cu(H_2O)_6]^{2+} + 2\ en$ ⇌	$[Cu(en)_2(H_2O)_2]^{2+} + 4H_2O$	$lg\beta_2 = 19.6$
$[Cu(H_2O)_6]^{2+} + trien$ ⇌	$[Cu(trien)(H_2O)_2]^{2+} + 4H_2O$	$lgK_1 = 20.1$
$[Cu(H_2O)_6]^{2+} + 2,3,2\text{-tet}$ ⇌	$[Cu(2,3,2\text{-tet})(H_2O)_2]^{2+} + 4H_2O$	$lg\beta_4 = 23.2$
$[Cu(H_2O)_6]^{2+} + cyclam$ ⇌	$[Cu(cyclam)(H_2O)_2]^{2+} + 4H_2O$	$lg\beta_4 = 26.5$

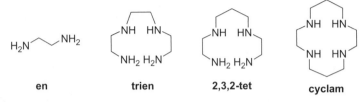

en   **trien**   **2,3,2-tet**   **cyclam**

(b)

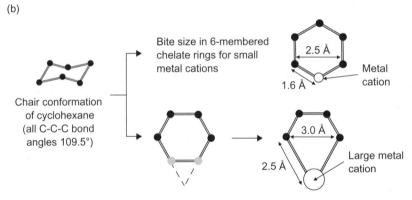

Chair conformation of cyclohexane (all C–C–C bond angles 109.5°)

Bite size in 6-membered chelate rings for small metal cations

2.5 Å

Metal cation

1.6 Å

3.0 Å

Large metal cation

2.5 Å

Bite size in 5-membered chelate rings for large metal cations

Figure 1.4 (A) Binding constants of nitrogen-containing ligands of varying denticities to Cu^{2+} (Adapted from R. E. Hancock & A. E. Martell, *Comments Inorg. Chem.*, 1988, **6**, 237, with permission from Taylor and Francis.); (B) greater stability of five-membered chelate rings for large metal cation complexes compared to six-membered rings.

chelate ring is identical to the chair form of cyclohexane, whose C–C–C bond angles are optimal at 109.5°, giving rise to low ring strain (Figure 1.4B). In the case of the chelate ring, one carbon atom of cyclohexane is replaced with the metal cation, with the two adjacent ones being the coordinating atoms (e.g. N). A large metal cation, however, requires more space to bind. Imagine that two adjacent atoms on the six-membered ring are removed and replaced with a single atom, whilst retaining all the other bond angles at 109.5°. The resulting five-membered chelate ring now possesses a larger bite size of 3.0 Å between the chelating atoms (instead of 2.5 Å for the six-membered ring), with longer bonds with the metal now possible as well (2.5 Å instead of 1.6 Å for the six-membered ring). Hence, five-membered chelate rings are preferred for large metal cations due to greater availability of space to bind the cation, reducing the overall ring strain of the chelating bridge. At ring sizes larger than six, the chelate effect reduces in

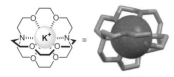

Figure 1.5 A crystal structure of a cryptand host forming a three-dimensional complex completely encapsulating a K$^+$ cation. Hydrogen atoms are omitted for clarity. Crystal structure refcode: CSD-BACCIS, first reported: R. Alberto *et. al.*, *J. Am. Chem. Soc.*, 2001, **123**, 3135.

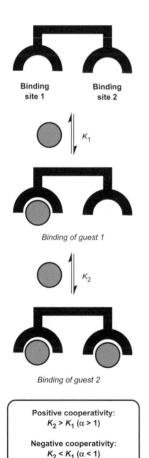

Figure 1.6 Schematic diagram showing positive, negative, and no binding cooperativity on a receptor with two guest binding sites.

magnitude due to the greater conformational entropy of the chelating chain, reducing the probability of ring-closing.

Closing the gap in 2,3,2-tet to form the macrocyclic cyclam ligand results in a further increase in complex stability by more than three orders of magnitude. This remarkable effect is known as the **macrocyclic effect** and also arises from a combination of factors. Compared with its acyclic analogues, macrocyclic ligands are less flexible and lose fewer degrees of freedom on metal complexation (in other words, binding is less entropically disfavoured). Macrocycles are also often less heavily solvated, and hence coordination is enthalpically favoured as less energy is needed for ligand desolvation before binding can take place. While macrocycles such as cyclam bind the guest in primarily two dimensions, adding a third strap around the macrocycle containing additional chelating atoms now completely encloses the metal cation in a three-dimensional cage (Figure 1.5). The resulting cryptand boasts the highest guest binding stabilities of all, conferred by the increased dimensionality of ligation, by the **cryptate effect**. Unsurprisingly, the rates of guest decomplexation decrease in the order of chelate > macrocyclic > cryptate complexes (most kinetically inert).

Cooperativity

The chelate, macrocyclic, and cryptate effects surveyed above are examples where the coordination of one binding motif (e.g. amines) to the metal centre positively reinforces the coordination of subsequent groups. Indeed, situations involving multiple binding to a host molecule, where the binding at one site affects the guest affinity at another site(s), are very common in natural enzymes and in synthetic supramolecular systems. This phenomenon is known as **cooperativity**, and although a comprehensive treatment lies beyond the scope of this primer, it can be simplistically classified into three scenarios. In positive cooperativity (Figure 1.6), binding at one site (K_1) increases the affinity at a second site (K_2), such that $K_2/K_1 > 1$, and are seen with the chelate, macrocyclic, and cryptate effects. In nature, the binding of O_2 to haemoglobin provides another good example of positive cooperativity. This metallo-protein comprises of four oxygen-binding sites, each capable of binding to one O_2 molecule. When one O_2 molecule binds, changes in the protein structure allows each subsequent O_2 molecule to be bound more strongly than the last. Conversely in negative cooperativity, the first binding event disfavours complexation of the second guest ($K_2/K_1 < 1$). This is observed in allosteric inhibition of enzymatic activity, where the binding of a molecule to the enzyme induces changes to the structure of its active site to reduce its affinity for the target substrate, and is important for homeostatic regulation. Finally, when the binding of one guest occurs completely independently of the subsequent ones (i.e. binding at one site does not affect the affinities of subsequent events) such that $K_2/K_1 = 1$, no cooperativity occurs.

Non-covalent interactions: the supramolecular chemist's toolkit

Unlike synthetic chemists who join atoms and molecules together with permanent covalent bonds, the supramolecular chemist links molecules together using *reversible* non-covalent interactions. These can be broadly classified as follows:

(a) electrostatics (ion–ion, ion–dipole and dipole–dipole)

(b) hydrogen bonding

(c) π-interactions (π–π stacking, cation–π and anion–π)

(d) van der Waals forces (dispersion and inductive forces)

(e) halogen bonding

(f) hydrophobic or solvophobic effects.

These interactions vary greatly in strength. For example, the strengths of dispersion forces are usually < 2 kJ mol^{-1}, while hydrogen bonds can range from ca. 10 kJ mol^{-1} (N–H$\cdots$:N in ammonia) to 160 kJ mol^{-1} (F–H$\cdots$:F in HF$_2^-$). Although individual supramolecular interactions are weaker than covalent bonds (ca. 350 kJ mol^{-1} for C–C bond), supramolecular chemists often use combinations of many interactions to achieve strong guest binding. In addition, the directionality of some interactions such as hydrogen and halogen bonding has been very useful to optimize the geometry and spatial orientations of the interacting species.

- **Electrostatics**: the Coulombic attraction between ions of opposite charges is non-directional, and is arguably amongst the strongest non-covalent interactions (up to 250 kJ mol^{-1}). Ion–ion interactions have a smaller dependence ($E \propto r^{-1}$) than ion–dipole ($E \propto r^{-2}$) and dipole–dipole interactions ($E \propto r^{-3}$), although the latter two interactions are more directional and require proper alignment to maximize binding strength (Figure 1.7). Due to their strength, electrostatics are commonly employed to bind charged guests (see Chapter 2), and can hold guests in place even in highly competitive solvents such as water (see Section 1.3).

- **Hydrogen bonding**: when a hydrogen atom is bonded to an atom more electronegative than itself, the resulting electron-deficient hydrogen atom is able to interact with anions or lone pairs on electron rich atoms. Hydrogen bonds vary greatly in strength, depending on the atom the hydrogen atom is bonded to: those bonded to fluorine, oxygen, and nitrogen are strong hydrogen bond donors, while those bonded to electron-deficient carbon atoms are weaker, though this can be compensated for by having a large number of convergent interactions (Figure 1.8A). Although hydrogen bonds show a wide range of bond lengths and bond angles,

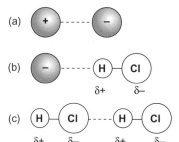

Figure 1.7 Electrostatic interactions: (A) ion–ion; (B) ion–dipole and (C) dipole–dipole.

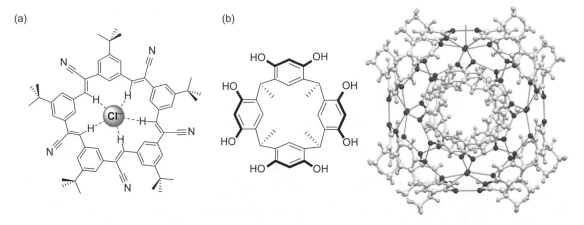

Figure 1.8 (A) A charge-neutral campestarene macrocycle for binding chloride exclusively utilizing C–H hydrogen bonds from electron-deficient carbon atoms. (B) Hydrogen-bonding mediated self-assembly of a spherical capsule of resorcinarene macrocycles held by 60 hydrogen bonds. (Crystal structure refcode: CSD-LIWQIR, first reported: L.G. MacGillivray and J.L. Atwood, *Nature*, 1997, **389**, 469)

their strength is directionally dependent and is greatest when the bond angles are approximately linear (180°). Due to its ubiquity in a large number of functional groups (e.g. amides, ureas, hydroxyls, etc.), and the ease with which it can be introduced at precise positions on molecules, hydrogen bonding is the most widely used supramolecular interaction found in nature and in designed supramolecular systems for host–guest recognition and self-assembly. An impressive example of the latter can be seen in the spherical molecular assembly of six resorcinarene macrocycles held together by 60 linear hydrogen bonds (Figure 1.8B). Hydrogen bonds are widely exploited for binding anions (Chapter 2), neutral guests (Chapter 3), and molecular self-assembly (Chapter 4).

- **π-interactions**: these occur between systems containing aromatic rings with other aromatic guests or ions, and are currently thought to be electrostatic in nature. π–π stacking occurs when two aromatic rings interact attractively with each other either in a face-to-face, offset parallel, or edge-to-face orientation (Figure 1.9A), and is an important stabilizing interaction in complex molecular architectures such as DNA and mechanically interlocked molecules (e.g. rotaxanes and catenanes in Chapter 5). Cation–π and anion–π interactions occur when the ions interact with electron rich and electron-deficient aromatic rings respectively (Figure 1.9B and C). Although cation–π interactions are recognized as important contributors to cation binding in nature (acetylcholine···tryptophan interactions) and for artificial hosts for cationic aromatic molecules, the roles played by anion–π interactions are only more recently recognized and exploited in anion binding host systems (Chapter 2).

- **van der Waals (vdW) forces**: Otherwise known as induced dipole–induced dipole interactions, vdW forces are very weak (< 2 kJ mol^{-1}) and are attractive interactions between instantaneous dipoles which form in the electron clouds of molecules. These interactions are very general in nature and are found between every pair of interacting molecules, hence they arguably make up small components of every supramolecular interaction. Despite their weakness, they are known to influence the supramolecular orientation of reacting species, sometimes favouring the formation of certain (unexpected) products over others. They are also known to enthalpically stabilize binding of guests inside restricted

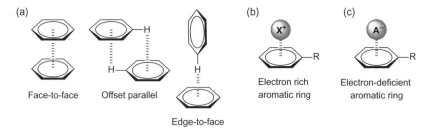

Figure 1.9 (A) Orientations of aromatic rings during π–π stacking interactions; (B) cation–π and (C) anion–π interactions.

molecular cavities, such as that between tetrahydrofuran (THF) and a calixarene derivative (Figure 1.10).

- **Halogen bonding**: Halogen bonding occurs when an electron-deficient heavy halogen atom (Br/ I) interacts with a Lewis base, which can be an electron rich atom or an anion. Despite being somewhat counter-intuitive that halogens, which are themselves electronegative, can act as Lewis acids, the conundrum can be resolved by considering the electron distribution on the halogen atom. When bonded to an electron-withdrawing group (E), the electron density on the halogen atom (X) is anisotropically re-distributed such that an electron rich 'equatorial belt' is formed, accompanied by an electron-deficient region immediately collinear with the E-X bond (Figure 1.11). This region is termed a 'sigma (σ)-hole' and is responsible for all attractive interactions with Lewis bases and accounts for the very stringent linearity (> 170°) of the halogen bond, much more so than hydrogen bonding. The strength of halogen bonding is comparable to hydrogen bonding (10 to 200 kJ mol^{-1}) and increases with increasing polarizability of the donor halogen atom, resulting in iodine being a stronger donor than bromine and chlorine. Having more electron-deficient groups bonded to the halogen atom also gives rise to stronger halogen bonds. Owing to its strength and linearity, it has been widely exploited in crystal engineering, molecular self-assembly to form nanocapsules (Chapter 4), and anion binding and sensing (Chapter 3).

- **Hydrophobic effect**: This is the driving force for the association of hydrophobic molecules in aqueous solution and is responsible for phenomena such as oil–water phase separation, formation of lipid bilayer membranes, and plays important roles in self-assembly of DNA and protein folding in nature. It is driven fundamentally by the tendency to minimize the energetically unfavourable interactions between polar water molecules and apolar hydrophobic surfaces. As shown in Figure 1.12, the structured array of water molecules around a hydrophobic surface (e.g. host binding cavity) are released upon interaction with an apolar substrate, increasing the disorder and entropy of the system. The release of these water molecules also allows them to form stronger hydrogen bonding interactions with other water molecules in the bulk phase than those with the apolar solute, enthalpically driving the process as well. Hydrophobic interactions are central to the formation of inclusion complexes between molecules possessing hydrophobic cavities (e.g. cyclodextrins, calixarenes, cucurbiturils) and hydrophobic guests which can be neutral

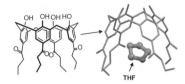

Figure 1.10 X-ray crystal structure of a vdW nanocapsular complex between THF and a calixarene derivative. Hydrogen atoms omitted for clarity. Crystal structure refcode: CSD-HEXSIN, first published: G. S. Ananchenko *et. al.*, *Cryst. Growth Des.*, 2006, **6**, 2141.

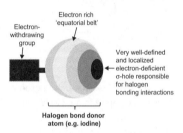

Figure 1.11 Distribution of electron density on a halogen bond donor atom (i.e. iodine) showing the location of the electron-deficient sigma hole responsible for halogen bonding interactions, as well as the electron rich equatorial belt.

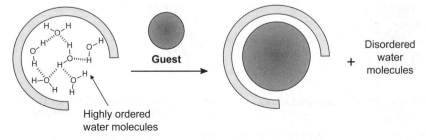

Figure 1.12 The origin of the hydrophobic effect.

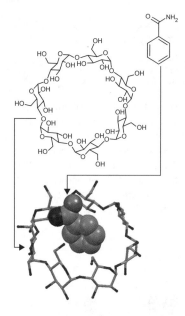

Figure 1.13 A crystal structure of an inclusion complex formed between β-cyclodextrin and benzamide. Hydrogen atoms omitted for clarity. (Crystal structure refcode: CSD-DEWCOY, first published: E. J. Wang *et. al., Carbohydr. Res.*, 2007, **342**, 767.)

(e.g. adamantane) or ionic (poorly hydrated ions), and can drive self-assembly processes in water (Chapter 4). Figure 1.13 shows an example of an inclusion complex where a molecule of benzamide is 'included' within the hydrophobic cavity of β-cyclodextrin. Alongside the hydrophobic effect, ***solvophobic forces*** are used to describe analogous behaviour in organic solvents.

1.4 Solvent effects

We have so far focused almost exclusively on the interactions between individual molecular components designed to bind with each other, without considering the influences of another very important component of the system—the solvent. The importance of the solvent in molecular recognition cannot be overstated, especially since it is present in so much greater excess than any of the interacting supramolecular species. Most of the time, the solvent will solvate the host, guest, and complexed species and can even interact strongly with itself. During host–guest binding, desolvation of the host and guest must occur to some extent (Figure 1.14). Although desolvation is enthalpically disfavoured due to the energetic demands required to break the bonds with the solvent, this process also gives a net increase in the number of free solvent molecules, which is entropically favoured. Furthermore, restoration of solvent–solvent interactions may release energy as well (favourable enthalpy). Taken together, the complex interplay of solvent interactions can exert dramatic influences on the thermodynamic balance and resulting host–guest equilibria, completely undermining rational host design if not properly accounted for. The hydrophobic effect, as we have seen in the preceding section, is a prime example of the solvent's (in this case water) important positive influence in driving host–guest recognition and binding.

Other than merely associating with the interacting molecular species, the size and shape of the solvent can also influence host–guest binding. This can be clearly seen when considering the binding of the 3D cyclophane in Figure 1.15 with imidazole in chlorinated hydrocarbons and ethers. Although the ethers can solvate the host and guest better by hydrogen bonding than the chlorinated hydrocarbons, resulting in generally weaker binding affinities, more sterically hindered members of each class of solvent give stronger imidazole binding. The effects are dramatic, increasing 3 orders of magnitude from $240\ M^{-1}$ in CH_2Cl_2 to $128\,000\ M^{-1}$ in $CHCl_2CHCl_2$, and from $29\ M^{-1}$ in THF to $1067\ M^{-1}$

Equation 1.1:

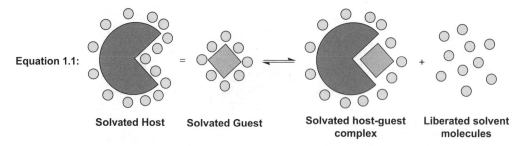

Solvated Host   Solvated Guest   Solvated host–guest complex   Liberated solvent molecules

Figure 1.14 Changes in solvation of the host, guest, and bound complex during host–guest recognition (schematic showing solvation of host and guest and release of solvent on binding, Equation 1.1).

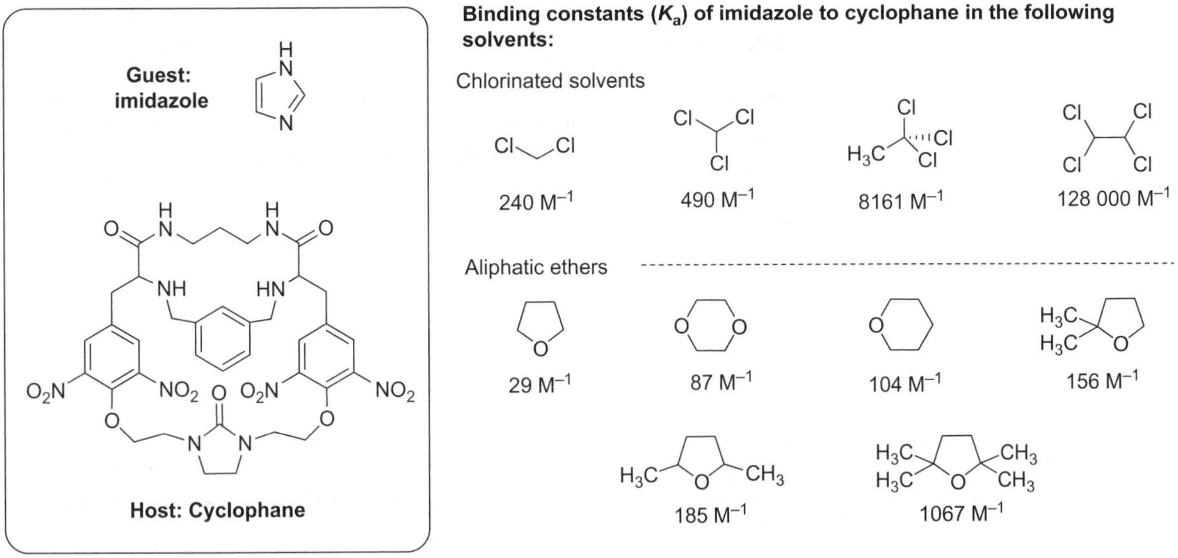

Guest: imidazole

Host: Cyclophane

Binding constants (K_a) of imidazole to cyclophane in the following solvents:

Chlorinated solvents

240 M^{-1} 490 M^{-1} 8161 M^{-1} 128 000 M^{-1}

Aliphatic ethers

29 M^{-1} 87 M^{-1} 104 M^{-1} 156 M^{-1}

185 M^{-1} 1067 M^{-1}

Figure 1.15 Effects of solvent size on imidazole binding to a cyclophane host. Adapted with permission from K. T. Chapman & W. C. Still, *J. Am. Chem. Soc.*, 1989, **111**, 3075. Copyright (1989) American Chemical Society.

in 2,2,5,5-(CH$_3$)$_4$-THF. The very strong imidazole binding in CHCl$_2$CHCl$_2$, with the largest-sized solvent molecules, is attributed to the lack of competitors for guest binding as the solvent molecules are unable to penetrate the cyclophane cavity.

Other than molecular size, what other properties of solvents influence host-guest binding equilibria? The most frequently considered parameters are the solvent *dielectric constant* and its donor–acceptor properties.

The dielectric constant (ε) of a solvent is a measure of the solvent's bulk polarity and can be considered its ability to screen electrostatic interactions. Solvents with individual molecules possessing high dipole moments generally have larger dielectric constants, and can therefore interact more strongly with charged species and dipoles. This effectively reduces the mutual influence of these ions and dipoles to screen them from one another. As a result, ion-pairs, which are very strongly held together in solvents of low polarity (Equation 1.2), become drastically weakened in solvents of high dielectric constant (e.g. water), leading to their dissociation.

The solvent's donor–acceptor numbers are empirical measures of its ability to donate or accept electron pairs. A good donor solvent has a propensity to donate electron pairs, and can readily solvate cations and accept a hydrogen bond. Conversely, a good acceptor solvent is also a good hydrogen bond donor and can effectively solvate anions by interacting with their lone pairs. The Gutmann donor and acceptor scale (Table 1.1) quantifies these experimental effects and serves as a good gauge of the relative host–guest binding strength expected in different solvents. A good donor and acceptor solvent, such as water (Figure 1.16), is able to interact with charged or highly polar hosts and guests (including those which rely heavily on hydrogen bonding for binding) to solvate them very effectively, pulling the equilibrium in Equation 1.1 (Figure 1.14) to the left and reducing the binding affinities. These solvents, which are

The electric potential energy between two ions constituting an ion-pair is given by Equation 1.2, where Q$_1$ and Q$_2$ are the ion charges, ε is the solvent dielectric constant and r is the distance between the ions.

Equation 1.2: $E = -\dfrac{Q_1 Q_2}{4\pi\varepsilon r}$

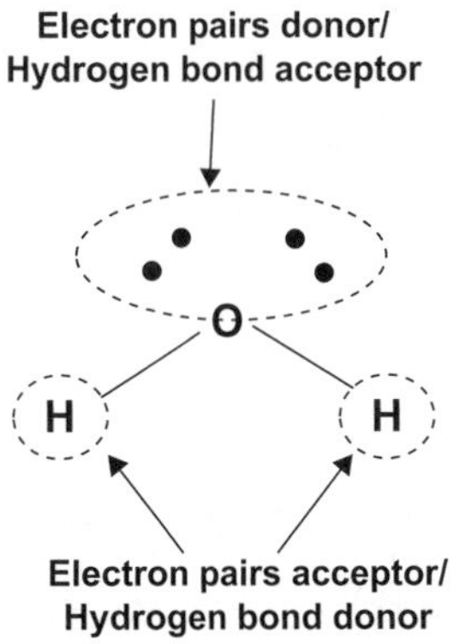

Electron pairs donor/ Hydrogen bond acceptor

Electron pairs acceptor/ Hydrogen bond donor

Figure 1.16 Origin of water's good donor–acceptor properties.

Table 1.1 The Gutmann donor and acceptor numbers and dielectric constants of common solvents. Reprinted from Electrochim Acta, V. Gutmann, *Empirical Parameters for Donor and Acceptor Properties of Solvents*, 661–670, 1976, with permission from Elsevier.

	Donor number (H bond acceptor)	Acceptor number (H bond donor)	Dielectric constant ε
H_2O	18.0	54.8	80.1
DMSO	29.8	19.3	46.7
DMF	26.6	16.0	36.7
Acetonitrile	18.9	14.1	36.6
Methanol	19.0	41.5	33.0
Ethyl acetate	17.1	9.3	6.0
THF	20.0	8.0	7.5
Dichloromethane	1.0	20.4	9.1
Diethyl ether	19.2	3.9	4.3
Hexane	no donor atoms	0.0	1.9

able to disrupt complex formation and compete with the guest for binding to the host, are known as *competitive solvents*.

Due to these factors, achieving strong and selective molecular recognition in water still remains one of the greatest challenges facing supramolecular chemists today. This is especially so when charged guests are involved, due to water's excellent solvating and charge screening properties ($\varepsilon = 80$, which is the largest amongst common solvents in Table 1.1). The detailed considerations are beyond the scope of this book, but the strategies applied depend on the nature of the guest to be bound. We have already seen how apolar guests in water can be strongly bound in water by capitalizing on the hydrophobic effect. However, the more challenging binding of hydrophilic guests, especially small charge-dense ions such as Li^+, Na^+, F^-, and Cl^- often require the use of highly charged hosts to partially mitigate water's excellent charge-screening properties. In addition, multiple convergent interactions (e.g. hydrogen bonding) within highly preorganized and complementary three-dimensional host binding cavities are usually needed. Regardless, a good understanding of the solvent properties and their roles in molecular recognition greatly aids the supramolecular chemist in designing appropriate hosts to maximize the strength of guest binding for any desired application.

1.5 How are supramolecular interactions and complexes studied?

Having carefully designed the host molecule for binding of the desired guest in the solvent of interest, the supramolecular chemist now needs to actually study the supramolecular host–guest complex formation itself. The following questions are frequently asked:

- What is the molecular structure of the host-guest associated complex?
- the kinetics: how rapidly is the host–guest associated complex formed?

- the thermodynamics: how strong are the interactions and how stable is the host–guest complex?

The ability to characterize supramolecular complexes accurately can be achieved using a range of techniques. We provide here an introduction to the most common and powerful methods frequently employed to study supramolecular host–guest complexation. Amongst them, nuclear magnetic resonance (NMR) techniques are the first points of reference, providing information to address all the essential points highlighted above.

Structural information

NMR: The utility of this technique arises from the abundance of nuclei with appropriate nuclear spin (I = ½) detectable by NMR spectroscopy, which include some of the most common atoms in organic molecules (^{1}H, ^{13}C, ^{19}F, ^{31}P). The frequency of atomic resonance is highly dependent on the local electronic environment around the atom, and hence NMR offers a unique ability to probe any perturbations to the atom's electronic environment which can occur upon interaction with another molecule. To study binding, NMR **titration** experiments are commonly performed. Typically, the host molecule's NMR spectrum in a deuterated solvent is first recorded, before the addition of small aliquots of guest dissolved in the same solvent. When binding occurs, the guest perturbs the electronic environment around nuclei closest to the binding site more than those located further away, enabling information about the location of the site of host binding to be elucidated. Further confirmation can be obtained by *Nuclear Overhauser Effect Spectroscopy* (NOESY), which correlates proton resonances near to each other spatially, even if the interacting nuclei are on separate molecules. These often manifest as cross-peaks on a 2D spectrum. While NOESY is effective for small or large molecules, species with molecular masses of around 1000 Daltons cannot be studied using this technique, as their frequency of molecular tumbling in solution causes negligible NOE signals even for protons which are very near to each other. As a significant number of supramolecular host–guest complexes fall within this size regime, a different type of NOE experiment, called *Rotating Frame Overhauser Enhancement Spectroscopy* (ROESY), has been devised to overcome this limitation. Similar to NOESY, nuclei in close proximity to each other correlate to give detectable cross-peaks. Taken together, these NMR techniques can allow the geometry of guest binding to be determined.

    In recent years, diffusion-ordered spectroscopy (DOSY) has become increasingly popular as an NMR technique to study host–guest complexes in solution. This experiment allows the diffusion coefficient of the molecule or complex to be determined (Figure 1.17A), which is inversely proportional to its hydrodynamic radius (r) (i.e. a larger molecule will diffuse at a slower rate in solution than a small one, giving a smaller diffusion coefficient), as shown in Equation 1.3.[2] If a significant molecular

---

[2] The diffusion coefficient D is related to the molecule's hydrodynamic radius r by the Stokes–Einstein equation:

Equation 1.3: $D = \dfrac{kT}{6\pi\eta r}$

where k = Boltzmann's constant, T = temperature and η = dynamic viscosity

(a)   (b)

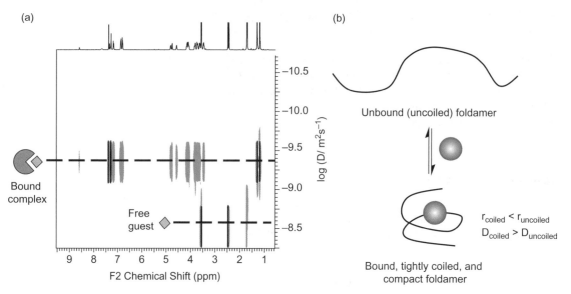

Figure 1.17 The use of DOSY to study host–guest binding with (A) significant changes in dimensions of complexes compared with individual guest molecules (authors' own data) (B) significant conformational changes when a foldamer wraps around a guest to adopt a compact conformation.

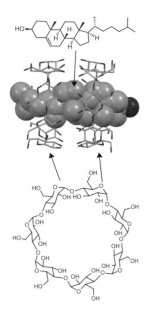

Figure 1.18 Crystal structure of a cholesterol molecule encapsulated by two β-cyclodextrin host species. Hydrogen atoms omitted for clarity. Crystal structure refcode: CSD-KEXQUC, first published: E. Christoforides *et. al.*, *Beilstein J. Org. Chem.*, 2018, **14**, 838.

size difference occurs during binding, for example when two or more molecules of similar size bind and form a complex, this can be readily detected, which also gives an indication of the stoichiometry of binding. This technique is also very useful to probe changes in molecular conformation during binding. As shown in Figure 1.17B, when a long foldamer molecule wraps around a small guest species (e.g. Cl⁻), its more compact conformation can be detected by an increase in diffusion coefficient as it can now diffuse faster in solution.

Crystallography: This technique provides the gold standard of characterizing supramolecular complexes, providing solid-state information on the geometry of host–guest binding, stoichiometry of binding, conformation of binding site, as well as the types of interactions which hold the complex together (e.g. HB, XB, aromatic π–π stacking). This can be illustrated by considering the crystal structure of the inclusion complex formed between β-cyclodextrin and cholesterol shown in Figure 1.18. This structure unambiguously shows that one cholesterol guest is encapsulated within two β-cyclodextrin host molecules, of which the latter are orientated such that the wider opening of each cyclodextrin faces each other. Although binding is driven largely by the hydrophobic effect, numerous hydrogen bonds between the encapsulated cholesterol molecule and the β-cyclodextrins also contribute in holding the complex together. Compared with NMR techniques, which probe the complex formation *in solution*, crystallography only provides information of the complex in the solid state, which may not necessarily be identical to that in the solution-phase. This difference arises due to solid-state packing effects present in the crystal, which are not present in solution, and can potentially distort the molecular structure. Despite its usefulness, crystallography can be hampered by difficulties in growing single crystals of supramolecular complexes which are of good enough quality for unambiguous structural determination. This problem is usually exacerbated with larger and more complex host–guest complexes, such as those involving rotaxanes (Chapter 5). Fortunately, supramolecular chemists can nowadays turn to computational simulations to provide insights on the structure and geometry of the host–guest complexes when crystallography proves too challenging.

Computational simulations: Modern computers can simulate the behaviour of molecules and their interactions to such a degree of reliability and accuracy that computational approaches are now routine in the study of catalysis, drug interactions, and polymers, to name a few. While a full treatment of this subject is beyond the scope of this primer, numerous techniques are also applied in supramolecular chemistry to study the optimum host–guest binding conformation. The simplest approach is based on *molecular mechanics* (MM), which treats atoms as solid hard spheres linked together by elastic bonds behaving like harmonic oscillators. Conversely, the most complex (and accurate) *ab initio* quantum mechanical methods attempt to find approximate solutions to the Schrödinger equation involving many electrons. As these methods are extremely demanding computationally, approaches such as the commonly used *density functional theory* (DFT) have been devised to simplify the calculations, based on the theorem that a molecule's energy minimum can be determined from functionals (i.e. functions of a function) of its spatially dependent electron density. Intermediate in complexity and computational demands are *semi-empirical* methods, which consider only the valence electrons in the quantum mechanical treatment. Ultimately, the choice of computational technique to use depends heavily on the size (i.e. number of atoms) of the supramolecular system considered, the type of information required, and the computing resources available. In a typical experiment, the host–guest's structure may be first generated from programs such as ChemDraw™ and then subjected to MM to generate a first approximation of the equilibrium geometry. More accurate methods such as DFT can then be used to interrogate energy-minimized binding conformations in the gas phase. To have a better understanding of how the host and guest interact in a solvent medium, *molecular dynamics* (MD) simulations can be performed using the prior DFT-optimized host–guest system and a large excess of associated solvent molecules. MD can simulate the physical movements of interacting atoms in the host/guest/solvent ensemble. Although less accurate than DFT, these solvent-containing ensembles can contain thousands of atoms, and are usually too computationally and time-demanding for detailed DFT simulations to be performed on all molecules present. Nonetheless, MD simulations still allow valuable information to be gleaned such as changes in host–guest binding conformations, binding selectivity trends, as well as the types and number of interactions involved in the simulated solvent medium.

Other techniques: Optical techniques such as UV-Vis and fluorescence spectroscopy are useful for probing the binding between molecules which possess conjugated π-electron systems or transition metals, as binding in the vicinity of these photoactive units often perturbs the electronic transitions which give rise to these spectra. Similarly, electrochemical voltammetric techniques such as cyclic voltammetry (CV) or square wave voltammetry (SWV) can be used to detect binding which can change the redox potential of the interacting species, especially so for charged guests binding near to the redox-active unit (e.g. ferrocene). Circular dichroism (CD) spectroscopy is particularly useful for studying conformational changes of chiral molecules during binding, and is especially informative when biomolecules such as DNA or peptides are involved, as specific secondary structures (e.g. α-helices) give well-known diagnostic CD spectra. In addition, infrared (IR) spectroscopy can often provide useful information on non-covalent interactions by perturbations to the stretching frequencies of certain diagnostic bonds. For instance, hydrogen bonding between an amine (RNH_2) donor and a carbonyl (O=C) acceptor often weakens both the N–H and C=O bonds, resulting in shifts of their absorbance bands towards lower wavenumbers.

Techniques for determining the aggregate mass of complexes are useful to determine the stoichiometry of host–guest binding. Mass spectrometry is nowadays a popular technique for this purpose, provided that the ionization methods used are mild (e.g. electrospray ionization (ESI) or matrix-assisted laser desorption ionization (MALDI)) such that the host–guest complex does not dissociate within the spectrometer before it reaches the detector. For large aggregates, size exclusion chromatography (SEC), on which the largest species migrate the fastest, can give information on the molecular mass as well.

Kinetic information

The formation of a host–guest complex is a dynamic process involving the exchange between bound and unbound forms of the host at equilibrium. As different spectroscopic techniques operate on different timescales, the kinetics of diffusion-controlled binding cannot be probed using techniques which operate on extremely short timescales. For instance, the typical electronic transition lifetimes of around 10^{-15} s in optical spectroscopy (UV-vis or fluorescence) are much faster than diffusion-controlled processes, causing all binding events to show slow exchange kinetics in comparison. In this regard, the appropriate operating timescales of NMR spectroscopy can offer insights on the kinetics of the binding process.

If the binding is *kinetically slow* compared to the frequency separation of the free and bound host NMR resonances (i.e. lifetime of the complex > 10^{-2} s), the addition of guest to host causes a new set of resonance peaks to appear which correspond to the bound host (Figure 1.19A). Accordingly, addition of increasing quantities of guest

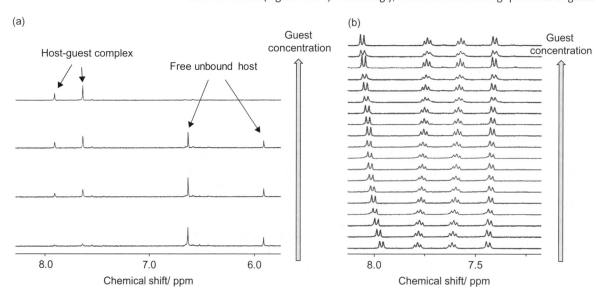

Figure 1.19 A series of NMR spectra showing (A) slow exchange kinetics between bound and free host species; (B) fast exchange kinetics on the NMR timescale (Authors' own data).

causes the resonances of the free host to decrease in intensity and that of the bound species to grow. In contrast, for a binding event kinetically fast compared to the NMR timescale, separate peaks corresponding to free and bound host are not seen. Instead, the host's NMR resonances are observed as an averaged peak whose frequency depends on the relative proportions of free and bound species (Figure 1.19B). Upon addition of more guest species, the host's time-averaged peak shifts continuously until the host is completely saturated.

Quantitative information about the rate of complexation can be obtained from NMR experiments. The kinetics of any dynamic process, such as host–guest binding, can be altered by simply changing the temperature. If a system showing slow exchange is heated, the individual peaks corresponding to free and bound host gradually broaden and then coalesce (i.e. merge) to form a single broad peak. The rate of complexation at the coalescence temperature can be determined using Equation 1.4.[3]

Thermodynamic information

In supramolecular chemistry, the strength of host–guest binding is one of the most important parameters that is quantified experimentally. Spectroscopic (NMR, UV-Vis, fluorescence) titration experiments allow determination of *binding* or *association constants*, often denoted as K_a.[4] This parameter is related to the Gibbs free energy of binding (ΔG), which is in turn dependent on the enthalpy (ΔH) and entropy (ΔS) of binding (Equations 1.5 and 1.6).[5]

Using NMR titrations as an example, the method of K_a determination depends on the kinetics of host–guest binding, and stems from determination of the relative concentrations of host and complex in equilibrium. For systems in slow exchange, integrating the proton resonances arising from the free and bound host gives their relative concentration, allowing K_a determination for a 1:1 stoichiometric complex as the concentrations of free host, guest, and bound complex are all known. For binding which is kinetically fast on the NMR timescale, simple peak integration cannot be carried out to determine concentrations. Instead, the shifting peak resonances can be plotted as a function of quantity of guest added to give a titration curve (Figure 1.20A). Its curvature contains all the necessary binding information, and is indicative of the host–guest ***binding affinity***. For instance, the curve labelled 1 in Figure 1.20A shows much stronger binding than curve 2, as only minimal changes to the chemical shifts are seen

---

[3] **Equation 1.4:** $k = \dfrac{\pi(\delta v)}{\sqrt{2}}$

k: rate of complexation at coalescence temperature
δv: frequency difference between the NMR peaks under slow exchange conditions.

[4] A related parameter which is frequently encountered, more often in biochemistry, is the dissociation constant K_d, which is the reciprocal of K_a (i.e. $K_d = \dfrac{1}{K_a}$).

[5] **Equation 1.5:** $\Delta G = -RT \ln K_a$, where R is the molar gas constant and T is the thermodynamic temperature (in Kelvins).
Equation 1.6: $\Delta G = \Delta H - T\Delta S$

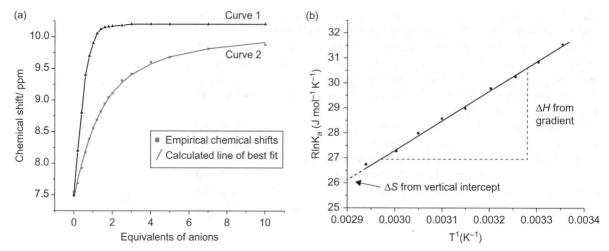

Figure 1.20 (A) Titration curves where the NMR chemical shift of resonance of the host molecule changes as a function of added guest concentration; (B) van't Hoff plot used to determine thermodynamic parameters ΔH and ΔS.

at >1.0 equivalents of guest (i.e. all host molecules are saturated).[6] This experimental data can be fitted to a theoretical model of the complexation using computer programmes, which can determine exact K_a values by employing non-linear least squares fitting algorithms. Although one cannot determine the values of ΔH and ΔS from these spectroscopic titrations directly, performing the titrations at different temperatures allow a van't Hoff plot to be obtained where $\ln K_a$ is plotted as a function of the reciprocal of the temperature T^{-1} (Figure 1.20B). From Equation 1.7,[7] ΔH and ΔS can be determined from the gradient and vertical intercept of this plot respectively, assuming that these values remain invariant with temperature.

The quantification of ΔH and ΔS can reveal valuable insights about the binding processes and the roles played by solvation. Fortunately, these can be determined more easily and accurately using calorimetry, which measures these parameters directly. *Isothermal titration calorimetry* (ITC) is a popular method where the temperature of a host solution is measured as a function of the quantity of guest added. The heat change reveals ΔH and ΔG, allowing ΔS to be easily calculated using Equation 1.6. Importantly, ITC can also reveal information about the stoichiometry of binding as well.

The range of techniques described above reveal different facets of host–guest binding, including guest selectivity, by comparing the binding constants with different potential guest molecules or competing species. This provides valuable information for the supramolecular chemist to refine the receptor's design to improve its selectivity, binding strength, or kinetics. Through multiple iterations, the receptor can become so finely tuned that it becomes suitable for technological applications, e.g. as selective sensors in functional devices.

[6] For NMR titrations, host–guest binding affinities of up to 10^4 M^{-1} may be accurately determined with initial host concentrations of around 1.0 mM. This is due to saturation of the binding curve with $K_a > 10^4$ M^{-1} (very small chemical shifts > 1.0 equivalents of guest). For instance, the binding isotherms from $K_a = 3 \times 10^4$ M^{-1} and $K_a = 5 \times 10^5$ M^{-1} are practically indistinguishable from each other.

[7] Substituting $\Delta G = -RT\ln K_a$ into Equation 1.6 gives $-RT\ln K_a = \Delta H - T\Delta S$. Rearranging this equation gives:

Equation 1.7: $R\ln K_a = \Delta S - \dfrac{\Delta H}{T}$

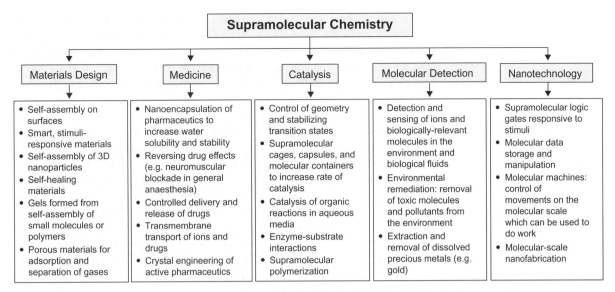

Figure 1.21 Flowchart showing some applications of supramolecular chemistry in different fields of modern research.

1.6 The modern importance of supramolecular chemistry

As we have mentioned before, supramolecular chemistry pervades many diverse fields of research, and numerous examples have found practical applications such as in drug delivery, molecular sensors, and even in odour control by the formation of inclusion complexes! The fundamental concepts of supramolecular molecular recognition and self-assembly are being used increasingly to design new generations of materials, explain phenomena such as materials behaviour and drug interactions, engineer new ways to extract and recycle precious metals, design new molecular sensors, and engineer new methods of manufacturing, just to name a few. Figure 1.21 summarizes some of the more exciting applications of supramolecular chemistry in various fields of research, and is by no means exhaustive.

In materials engineering, supramolecular chemistry governs the behaviour of liquid crystals, the spontaneous self-assembly of micelles, the long-range order of porous crystalline frameworks (such as metal–organic frameworks, or MOFs, which will be discussed in Chapters 3 and 4) as well as functional surfaces for nanodevices. Often, the applications of supramolecular materials are multi-disciplinary. A prime example of such materials are supramolecular hydrogels, which comprise of water immobilized within a solid-like three-dimensional self-assembled matrix, and can be assembled by virtually every interaction in the supramolecular toolkit: electrostatic interactions, hydrogen bonds, halogen bonds, hydrophobic interactions, and inclusion-complex formation. They are important in numerous biomedical applications, including but not restricted to three-dimensional cell cultures, tissue engineering, drug delivery, lubricants for joints, vitreous substitutes (for eye surgery), stabilizing agents for unstable proteins, and even as biomedical 'glues' for sticking wounds and sutures together to hasten recovery. In addition, multifunctional hydrogel sensors for discrete molecules, gases, biomolecules (e.g. proteins and nucleic acids), and microorganisms (e.g. bacteria, fungi, and viruses) are known. They have found applications in environ-mental remediation, being able to remove both soluble pollutants (e.g. heavy metal ions, toxic anions, dyes, pesticides) and non-soluble pollutants (e.g. oil spills) from water bodies. Even more intriguingly, hydrogels can be catalysts and non-conventional media for chemical reactions, and have even been applied in soft molecular robotics! Often, their responsiveness to external stimuli, such as light and heat, are governed by the supramolecular interactions holding the entire hydrogel network together. In the biomedical field, supramolecular chemistry often plays crucial roles in drug formula-tions to increase the bioavailability, water solubility, and biological stability of many active drug molecules (see Chapter 3). Molecular sensing (Chapters 2 and 3), the ability to selectively recognize, bind, and give a detectable output signal, is a fundamental consequence of host–guest supramolecular interactions. Nanomachinery and supra-molecular devices (discussed in Chapter 5), which form important facets of modern nanotechnology, have their roots firmly in supramolecular chemistry. Indeed, as we progress along each chapter, you will find many important applications of supramo-lecular chemistry, which further emphasizes its eminence as a central tenet of modern chemical research.

1.7 Further reading

Supramolecular chemistry is an inherently vast and interdisciplinary field, encompass-ing aspects of spectroscopy, organic, inorganic, and physical chemistry. The following references cover various aspects of supramolecular chemistry in much greater detail than the scope of this short book provides:

- General aspects of supramolecular chemistry: Steed, J. W. and Atwood, J. L., 2009, *Supramolecular Chemistry*, 2nd Edition, Chichester: Wiley.
- Determination of binding constants and stoichiometry of host–guest binding: *Chem. Soc. Rev.*, 2011, **40**, 1305–23.
- Applications of halogen bonding in supramolecular chemistry: *Chem. Rev.*, 2015, **115**, 7118–95.

1.8 Summary

- Supramolecular chemistry's two main themes of molecular host–guest recogni-tion and self-assembly are inspired significantly by natural biological examples.
- The concepts of preorganization and host–guest complementarity govern the design of many supramolecular synthons capable of interacting with one another via a wide range of non-covalent interactions.
- The design of receptors for target guest species needs to take into account the nature of the solvent in which the host–guest association takes place which can be exploited to influence guest affinity and selectivity significantly.

- Of the numerous techniques used to study supramolecular host–guest complexes, NMR is the most versatile and can offer a wealth of structural, kinetic, and thermodynamic information on recognition processes in solution.

- The concepts of supramolecular chemistry are essential to understand the interactions between molecules, and underpin many areas of modern research.

1.9 Exercises

The calculation of binding constants from titration experiments is one of the most important techniques a supramolecular chemist must master. The following exercises aim to give some insights on how to perform this in practice.

1.1   A host and a guest molecule are interacting in fast exchange on the NMR timescale in deuterated acetone. Table 1.2 below gives the changes in the chemical shifts of an aromatic proton on the host molecule with increasing quantities of guest. Initial [host] = 1.0 mM, T = 298 K.

Table 1.2   Changes in chemical shifts of an aromatic proton on the host at different host and guest concentrations.

Host concentration/M	Guest concentration/M	Aromatic proton chemical shift/ppm
0.001	1E-11*	7.9669
0.000996	0.000199	7.986
0.000992	0.000396	8.0047
0.000988	0.000592	8.0211
0.000984	0.000787	8.0354
0.000980	0.000980	8.0476
0.000976	0.00117	8.0588
0.000973	0.00136	8.0668
0.000969	0.00155	8.0722
0.000965	0.00174	8.0768
0.000962	0.00192	8.0807
0.000952	0.00238	8.0852
0.000943	0.00283	8.0887
0.000926	0.00370	8.093
0.000909	0.00455	8.0963
0.000877	0.00614	8.0992
0.000833	0.00833	8.1028
0.000769	0.0115	8.1049
0.000714	0.0142	8.1076

*note that the initial guest concentration provided just designates a very small non-zero value (Authors' own data).

There are available resources on the internet which offer an easy method of calculating binding constants by non-linear regression (see Further reading for more details). One such example is the website http://app.supramolecular. org/bindfit/, developed by Professor Pall Thordarsson at the University of New South Wales, Australia. First, create an input.excel file using the data above, by following the instructions provided on the website.

(a) Click on the 'NMR 1:1' tab and upload the input file you have just created. Set an initial estimate of K_a = 1000 M^{-1} and click on the 'Fit' tab. What is the binding constant and the error obtained? Try other initial estimates of K_a as well. Do they give the same value of K_a?

(b) Attempt fitting the same data to 'NMR 1:2' (this means a host: guest 1:2 stoichiometric ratio). Set 'Flavour' = 'None' and 'Method' = 'Nelder-Mead'. You should expect to obtain two binding constants, K_{11} being the first association between 1 host and 1 guest molecule, and K_{12} being the association of the second guest molecule onto the 1:1 bound complex. What are the values and errors of K_{11} and K_{12} obtained? Attempt fitting the data using 'Method' = 'L-BFGS-B'. What values of K_{11} and K_{12} are obtained?

(c) Based on the binding constants obtained in (a) and (b), what is the most likely binding stoichiometry between the host and guest? Why?

(d) Why is it important to report the temperature associated with a binding constant?

1.2   Two molecules are known to form a 1:1 stoichiometric host–guest complex in $CDCl_3$. When a solution containing 1.5 µmol of the host and 1.8 µmol of the guest in exactly 1.0 mL of $CDCl_3$ is analysed by ^1H NMR spectroscopy, the spectrum shows a peak for the host–guest complex and a separate peak for the unbound host. The area under the peak for the complex is exactly 3.6 times that of the unbound host. Calculate the K_a of the host–guest complex in $CDCl_3$ (T = 298 K).

1.3   Consider the following host–guest 1:2 stoichiometric binding equilibria:

$$H+G \underset{}{\overset{K_1}{\rightleftharpoons}} HG$$

$$HG+G \underset{}{\overset{K_2}{\rightleftharpoons}} HG_2$$

(a) Express the equilibrium constants K_1 and K_2 in terms of the concentrations of host (H), guest (G), 1:1 host–guest complex (HG), and 1:2 complex (HG_2).

(b) Show that $K_2 = \dfrac{[HG_2]}{K_1[H][G]^2}$

(c) Using the expressions obtained in (a) and (b), derive an expression for the overall stability constant β_{12} of the 1:2 host–guest binding in terms of concentrations of H, G, and HG_2.

For the answers to these exercises, visit the online resources which accompany this primer.

2 Binding of charged guests

2.1 Introduction

Although cations and anions represent opposite sides of the same coin, and cannot exist without each other, the development of their receptor chemistry follows very different timelines. The stimulus for the design of host molecules for the binding of cationic guest species arguably traces its genesis to 1967 with Charles J. Pedersen's reports of the synthesis of crown ether macrocycles capable of selectively binding alkali metal cations. Anion receptor chemistry followed shortly after in 1968, with Simmons' and Park's report of ammonium macrobicyclic receptors capable of halide anion binding under acidic conditions. In the following few years, an explosive growth in *cation* receptor chemistry resulted, with entire families of receptors developed to bind main group, transition metal, and lanthanide cations selectively (see Section 2.4). In contrast, other than a handful of reports in the 1970s, interest in *anion* receptor chemistry was rekindled only in the 1980s, as more chemists gradually realized the importance of this field. Notwithstanding its greater challenges compared to cation binding (Section 2.3), great strides in anion receptor chemistry have been achieved in the few decades since (Section 2.5). This chapter will introduce the supramolecular chemistry of binding charged guests, starting with cations (Section 2.4), then anions (Section 2.5), and finally, both cations and anions simultaneously (Section 2.6).

2.2 Why bind cations and anions?

The ubiquity of ions in the natural environment impact many biological and chemical processes. As a result, numerous naturally occurring receptors capable of selectively binding a wide variety of cations and anions have evolved. We have seen an example of a naturally occurring phosphate anion binding protein in Chapter 1 and how it offers selectivity over arsenate. In addition, many natural proteins capable of selectively binding anions (Cl$^-$, I$^-$, SO$_4^{2-}$, various carboxylates) are known. Biological processes such as photosynthesis (chlorophyll a), oxygen transport (haemoglobin), and metabolism (enzyme cofactors e.g. vitamin B$_{12}$) all rely on metal cations bound within macrocycles (Figure 2.1). Abiotic synthetic receptors for cations and anions have also been exploited in many important human activities. The most important applications of ion-receptor chemistry are summarized in Figure 2.2, and selected examples will be discussed in greater detail in Section 2.7.

2.1

2.2

2.3

Figure 2.1 Examples of naturally occurring cation binding macrocycles essential for biological functions: the porphyrin macrocycles of chlorophyll a (**2.1**) and haem (**2.2**) binding Mg^{2+} and Fe^{2+} respectively, as well as the corrin macrocycle for Co^{3+} binding in vitamin B$_{12}$ (**2.3**).

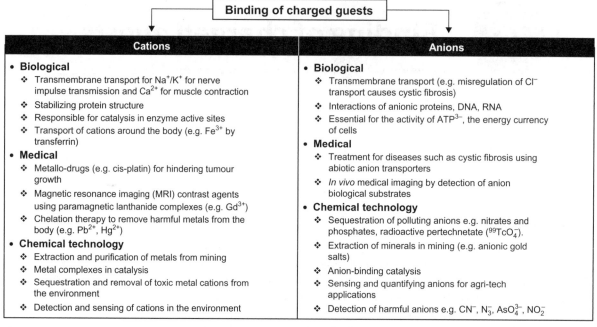

Figure 2.2 Key applications of cation and anion binding.

2.3 Design principles for cation and anion receptors

Superficially, one may be tempted to think of cations and anions as being similar, albeit with opposite charges. However, in reality, both classes of ions differ significantly in many ways, with important consequences for the design of their respective receptors. Before we consider the common design strategies for cation and anion receptors, it is important to first understand the fundamental differences between cations and anions, and how these factors make anion binding significantly more challenging than for cations.

Differences between cations and anions

Ionic radius: Anions are larger than their isoelectronic cations. For example, F^- has an approximately 30% larger effective ionic radius compared to the isoelectronic Na^+ (see Table 2.1). The resulting lower charge densities of anions result in weaker interactions with host molecules relying on electrostatic and/or ion–dipole interactions for binding.

Table 2.1 Comparison of the effective ionic radii and free energies of hydration of isoelectronic cations and anions. (Reproduced from Y. Marcus, *J. Chem. Soc., Faraday Trans.*, 1993, **89**, 713, with permission from the Royal Society of Chemistry).

Electronic configuration	Cation	Effective ionic radius (Å)	ΔG_{hyd} / kJ mol^{-1}	Anion	Effective ionic radius (Å)	ΔG_{hyd} / kJ mol^{-1}
[Ne]	Na^+	1.02	−365	F^-	1.31	−465
[Ar]	K^+	1.37	−295	Cl^-	1.81	−340
[Kr]	Rb^+	1.52	−275	Br^-	1.96	−315
[Xe]	Cs^+	1.67	−250	I^-	2.20	−275

Free energies of solvation: Compared with cations of similar size, anions often have higher free energies of solvation. For instance, both F^- and K^+ have similar ionic radii, however $\Delta G_{hyd}(F^-)$ of -465 kJ mol^{-1} is much larger than that of K^+ (-295 kJ mol^{-1}) (Table 2.1). This means that host molecules must be able to more effectively outcompete the surrounding solvent molecules for binding anions effectively. Thermodynamically, more energy needs to be expended to break the anion–solvent interactions, which is an endothermic process and reduces the favourable enthalpic contribution to binding.

Geometry: The majority of cation receptors are designed to bind spherical metal cations. In contrast, even simple inorganic anions occur with a variety of shapes and geometries (Figure 2.3). Biologically important organic anions such as carboxylates and oligophosphates (e.g. ATP^{3-}) have more complex shapes. These shapes need to be taken into account for receptor design in order to maximize the strength of complementary interactions between host and guest.

pH dependence: Whilst the vast majority of cations remain positively charged regardless of pH, many anions exist only within a narrow pH window. For instance, under highly acidic conditions, common oxoanions such as SO_4^{2-}, PO_4^{3-}, and CO_3^{2-} become protonated and lose their negative charge. These polyanions often also display multiple protonation equilibria, and slight changes in pH can change their charge, hydrophilicity, free energies of hydration, and degree of interaction with the respective host. Therefore, host molecules must be designed to function within the pH window of their target anions (Figure 2.4). This is especially tricky if ammonium-containing hosts are used, as their protonation states are themselves pH-dependent: the acidic pH values where the host molecules are protonated, and best able to interact with anions by electrostatic interactions and hydrogen bonding, may cause the anion guests to become neutral via protonation and hence insusceptible to binding.

Coordination preference: Cations, especially those of transition metals, can be coordinatively unsaturated, i.e. they possess vacant coordination sites for geometrically defined interactions/coordinate bond formation with Lewis basic atoms. This property can be exploited to engineer hosts which bind the cations tightly and selectively by maximizing the target cation's coordination directionality and preferences. On the other hand, anions are coordinatively saturated. Hence, they can only be bound via weaker interactions such as hydrogen bonding and halogen bonding.

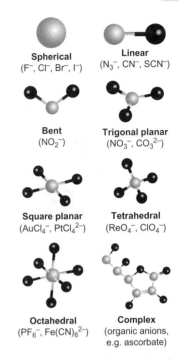

Spherical (F^-, Cl^-, Br^-, I^-)

Linear (N_3^-, CN^-, SCN^-)

Bent (NO_2^-)

Trigonal planar (NO_3^-, CO_3^{2-})

Square planar ($AuCl_4^-$, $PtCl_4^{2-}$)

Tetrahedral (ReO_4^-, ClO_4^-)

Octahedral (PF_6^-, $Fe(CN)_6^{2-}$)

Complex (organic anions, e.g. ascorbate)

Figure 2.3 Shapes and geometries of common anions.

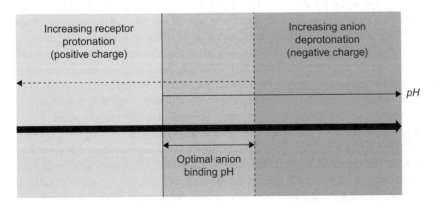

Increasing receptor protonation (positive charge)

Increasing anion deprotonation (negative charge)

pH

Optimal anion binding pH

Figure 2.4 Balance between host protonation and guest deprotonation for optimal binding.

(a)

(b)

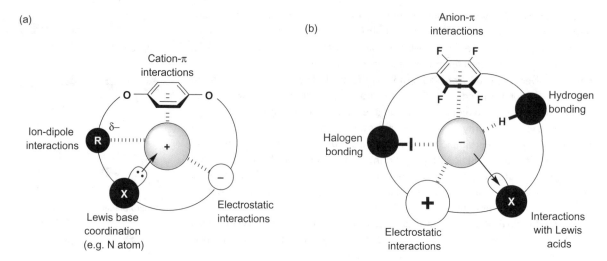

Figure 2.5 Common non-covalent interactions used in receptors to bind (A) cations and (B) anions.

Common design strategies

The differences between cations and anions highlighted in the previous section greatly impact the design of their receptors. The target ion's binding geometry, coordination preferences, and pH tolerance all need to be taken into account. Nonetheless, there are several design themes which are common and applicable to both cations and anions. Most obviously, the receptor must possess functional groups which *complement* the guest's charge: to bind cations, a choice of Lewis basic domains such as atomic lone pairs capable of donation, electron rich aromatic and anionic groups, need to be incorporated in the receptor design (Figure 2.5A). Conversely, anion binding receptors can be positively charged, possess strong hydrogen/halogen bond-donor groups, Lewis acids, and electron-deficient aromatic functionalities (Figure 2.5B). Obviously, a positively charged receptor would be ineffective for binding cations due to host–guest electrostatic repulsion!

To maximize binding affinity and selectivity, the receptor should also be highly preorganized with shape and size complementarity for the target guest ion. This is especially critical for anion receptors with their diverse shapes. For example, tripodal receptors with C_{3v} symmetry have been designed to bind the trigonal planar nitrate anion. As seen from Figure 2.6, each amide-containing 'arm' of receptor **2.4** is designed to form hydrogen bonds with each Lewis basic oxygen atom of the nitrate guest. In addition to geometric complementarity, the 'hard-soft' character and polarizability of the target ion should be taken into account as well. Generally, 'hard' ions with low polarizability such as Na^+ are best coordinated by similarly 'hard' atoms, such as oxygen. On the other hand, 'softer' and more polarizable metal cations with lower charge densities are bound more strongly using correspondingly 'softer' atoms such as nitrogen and sulfur. This is illustrated by the metal cation affinities of the series of crown ethers in Figure 2.7, where two oxygen atoms of [18]crown-6 (**2.5**) are replaced with either nitrogen (receptor **2.6**) or sulfur (receptor **2.7**). While [18]crown-6 binds the 'hard' K^+ cation more strongly than either receptors **2.6** and **2.7**, the trend is reversed for the softer Ag^+ and Pb^{2+} cations. Notably, both **2.6** and **2.7** bind both 'soft' cations more strongly than [18]crown-6.

Receptor design must also take into consideration the degree of ion solvation. The importance of guest solvation in host–guest recognition has been expounded

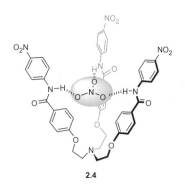

2.4

Figure 2.6 A C_{3v} tripodal receptor **2.4** for binding the trigonal planar nitrate anion.

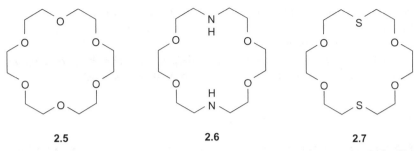

2.5   **2.6**   **2.7**

Figure 2.7 [18]crown-6 (**2.5**) and its derivatives **2.6** and **2.7** bearing 'softer' nitrogen and sulfur atoms respectively.

Hydrophilic *Strongly hydrated*	*Hydrophobic* *Weakly hydrated*
Cations : Guanidinium$^+$ > Ca^{2+} > Mg^{2+} > Li$^+$ > Na$^+$ > K$^+$ > NH$_4^+$ > (CH$_3$)$_4$N$^+$	
Anions : SO$_4^{2}$ > F$^-$ > H$_2$PO$_4^-$ > acetate$^-$ > Cl$^-$ > Br$^-$ > NO$_3^-$ > I$^-$ > ClO$_4^-$ > SCN$^-$	

Figure 2.8 The Hofmeister series for cations and anions.

upon in Chapter 1, and nowhere is this more important than in ion recognition. The wide variety of cations and anions inevitably result in a large range of hydrophilicities, and this can be powerfully exploited for selective ion binding. The ***Hofmeister series*** (Figure 2.8), originally established through studying the effects of salts on protein solubility, orders cations and anions by their increasing hydrophilicity and hence, greater degrees of aqueous solvation. Ions on the right of the series are more hydrophobic, and are best bound within similarly hydrophobic cavities which naturally exclude polar solvent molecules (e.g. water). On the other hand, hydrophilic ions on the left of the series are poorly compatible with hydrophobic cavities, and are more strongly bound in cavities which are similarly accessible to polar solvents.

Before taking a more in-depth survey of cation and anion binding receptors, an important point needs to be made regarding ion-pairing in solution. When performing binding studies, charged guests are always added as a salt with one or more counterions present. The choice of solvent necessitates a trade-off between degree of host/guest solvation and ion-pairing. While both host and guest may be poorly solvated in non-polar solvents and hence can potentially give strong binding affinities, the ion-pairing between the target ion and its counterion can also be quite strong, which reduces the 'availability' of the target ion for binding to the receptor. On the other hand, more polar solvents will weaken the strength of ion-pairing by charge-screening, whilst better solvating the host and guest and weakening their mutual interactions. Consequently, to reduce the impact of ion-pairing, charged-guest recognition in organic solvents is often performed using large charge-diffuse and poorly coordinating counterions. For example, tetrabutylammonium (TBA$^+$) is often used for anion-binding studies and hexafluorophosphate (PF$_6^-$) for cations. Often, the effects of ion-pairing are ignored in binding studies, which can be justified to some extent using these poorly coordinating counterions and reasonably polar solvent media. However, issues can arise when attempting to compare quantitative data obtained using different counterions, or methods (e.g. NMR and optical spectroscopy) which

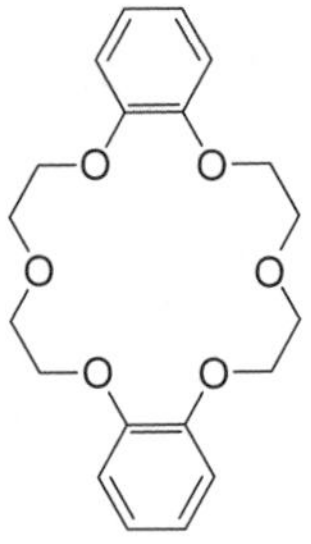

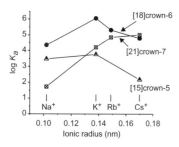

2.8

Figure 2.9 Dibenzo[18]crown-6 (Systematic name: 2,3,11,12-Dibenzo-1,4,7,10,13,16-hexaoxacyclooctadeca-2,11-diene).

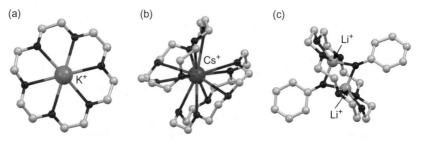

Figure 2.10 LogK for [15]crown-5, [18] crown-6 and [21]crown-7 for various alkali metal cations in methanol. No binding data for Rb⁺ with [15]crown-5 is available due to multiple host–guest binding equilibria. Adapted with permission from J. D. Lamb *et. al.*, *J. Am. Chem. Soc.*, 1980, **102**, 475. Copyright (1980), American Chemical Society.

function in different concentration regimes, especially in non-polar solvents. These caveats should be kept in mind when determining the optimal conditions for binding studies as well as interpreting the results.

The following sections of this chapter will describe the major classes of receptors which have been developed for binding charged guests in greater detail.

2.4 Cation binding receptors

From crown ethers to spherands: increasing selectivity and affinity

The story of how crown ethers were discovered is well-known. In 1962, while preparing multidentate ligands for the vanadyl (VO^{2+}) cation to prepare an olefin polymerization catalyst, Charles Pedersen isolated some white crystals in 0.4% yield. While most chemists would have ignored such a minor side product, Pedersen's innate curiosity led him to study these crystals further. Curiously, he found that the addition of sodium salts could greatly increase the crystals' solubility in methanol. Upon careful analysis, Pedersen determined the structure of the impurity he had isolated to be a cyclic polyether flanked by two catechol groups (Compound **2.8** in Figure 2.9). As the systematic name for this compound proved too cumbersome for use, this particular compound was named dibenzo[18]crown-6, where [18] represented the number of atoms making up the macrocycle and 6 being the number of oxygen atoms in the ring. This class of polyether macrocycles was dubbed '***crown ethers***' due to their crown-like conformation in the solid state. Subsequently, Pedersen determined that the enhanced solubility of the macrocycle in methanol in the presence of sodium cation came about from the cation being bound within the crown cavity, held by ion–dipole electrostatic interactions with the six oxygen donor atoms in the ring.

Generally, cations with more *optimal spatial fits* with the cavity of the crown ethers result in stronger binding. For instance, [18]crown-6 shows a distinct preference for binding the most optimally sized K⁺, whilst the larger [21]crown-7 binds Cs⁺ most strongly (Figure 2.10). This relationship between cation and cavity size from binding constants is also evident from the crystal structures of the complexes. As shown in Figure 2.11A, the optimal 1:1 host–guest binding between K⁺ and [18]crown-6 allows the cation to be bound in the middle of the macrocycle to maximize interactions with all six oxygen atoms. For larger cations such as Cs⁺, the cation is unable to completely fit into the macrocycle cavity, and hence forms a *sandwich complex* where the cation is 'perched' above each macrocyclic plane ring, whilst still bound to all oxygen atoms present

Crystal structure refcodes for Figure 2.11 and sources of first publication:
(A) CSD-FETJEU: S. A. Kotlyar *et al.*, *Acta Cryst.*, 2005, **E61**, m293.
(B) CSD-DUBCIM: S. B. Dawes *et. al.*, *J. Am. Chem. Soc.*, 1986, **108**, 3534.
(C) CSD-KEXDIA: K. A. Watson *et al.*, *Can. J. Chem.*, 1990, **68**, 1201.

(a)   **(b)**   **(c)**

Figure 2.11 Crystal structures of [18]crown-6 with (A) K⁺; (B) Cs⁺; and (C) Li⁺ (containing the phenoxide counteranions as well).

(Figure 2.11B). Conversely, the much smaller Li⁺ adopts an asymmetric binding mode where only some of the crown ether's oxygen atoms are involved in binding each cation, allowing the remaining oxygen atoms to bind to another Li⁺. In the example in Figure 2.11C, both Li⁺ cations are bridged and further stabilized by two phenoxide counteranions above and below the crown ether.

Even early on, the affinities of crown ethers for cations suggested practical applications as **phase transfer catalysts**. The solubility of crown ethers such as [18]crown-6 in organic solvents facilitated the movement of ions across an aqueous–organic solvent interface by binding to the cation. This has been exploited in the use of the strong oxidant, potassium permanganate ($KMnO_4$), for oxidizing organic molecules. To increase the solubility of $KMnO_4$ in organic solvents, catalytic quantities of [18]crown-6 crown ether are typically added into a biphasic reaction mixture of aqueous $KMnO_4$ and an organic phase containing the organic reactant of interest. The crown ether binds to K⁺, increasing the solubility of the salt in the organic phase. This allows the usually insoluble MnO_4^- oxidant to oxidize the desired reactant in the organic phase efficiently (Scheme 2.1).

Other than alkali metal cations, crown ethers are useful hosts for many other cations. Complexes with other main group metals, transition metals, and lanthanides have been isolated and characterized crystallographically (Figure 2.12A–C). We have also seen examples of how the substitution of oxygen with other heteroatoms (e.g. N and S) in the crown ether macrocycles can be exploited to bind 'softer' cations more strongly in the section on common design strategies earlier in this chapter (Section 2.3; Figure 2.7). Furthermore, complexes with organic cations are known. For example, the guanidinium cation is optimally bound within the cavity of the large crown ether benzo[27]crown-9 (Figure 2.12D), where all hydrogen atoms on the cation can form hydrogen bonds with oxygen atoms around the macrocycle.

Using crown ethers as the basic structural motifs, early generations of supramolecular chemists made structural modifications by appending different functionalities to enhance cation binding affinities. In 1980, George Gokel reported the first **lariat ethers**—crown

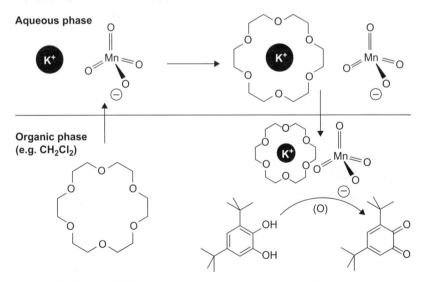

Scheme 2.1 The use of [18]crown-6 as a phase transfer catalyst for oxidizing 3,5-di-tertbutylcatechol in CH_2Cl_2 using aqueous $KMnO_4$.

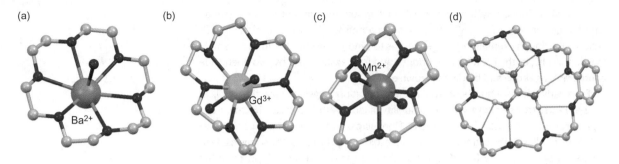

Figure 2.12 Crystal structures of [18]crown-6 with (A) barium and (B) gadolinium; (C) [15]crown-5 with Mn^{2+}; (D) guanidinium cation with benzo[27]crown-9. Hydrogen atoms are omitted for clarity except for the guanidinium cation in (D).

Crystal structure refcodes for Figure 2.12 and sources of first publication:
(A) CSD-FUYCEH: Y. Y. Wei *et. al., Acta Cryst.*, 1988, **C44**, 77.
(B) CSD-HAMZIE: R. D. Rogers *et. al., Inorg. Chem.*, 1993, **32**, 3451.
(C) CSD-RAYTIV: X. Hao *et. al., Cryst. Growth Des.*, 2005, **5**, 2225.
(D) CSD-BAYXII: W. H. M. Jos *et. al., J. Chem. Soc, Chem. Commun.*, 1982, 200.

ethers with pendant arms containing additional coordinating groups for cations. The pendant arms can be attached to a nitrogen atom on an aza-crown ether or to a carbon atom on the macrocycle to form *N-pivot* and *C-pivot* lariat crowns respectively, and can have multiple coordinating groups (Figure 2.13). Lariat ethers have greater affinities than their parent crown ethers for cations owing to the additional coordination(s) from the pendant arm; however, the cations remain kinetically labile due to the conformational flexibility of the arm. In addition, lariat ethers with two or three arms, respectively known as *bibracchial* and *tribracchial* lariat ethers, have also been synthesized.

Cryptands effectively brought cation binding into the third dimension (see Figure 1.5, Chapter 1 for a crystal structure). First reported by Jean-Marie Lehn in 1969, these caged molecules are able to bind cations much more strongly compared to crown ethers owing to the *cryptate effect* described in Section 1.3 in Chapter 1, attributed to a combination of entropic and enthalpic factors. For instance, [2.2.2]cryptand (see Figure 2.14) binds K^+ with a binding constant over four orders of magnitude larger than [18]crown-6 in methanol. Like crown ethers, the most stable cryptates (cation complexes of cryptands) correspond to an optimal size-match between the cation and the cryptand cavity. However, their more confined 3D cavities offer greater selectivity over other cations (Table 2.2).

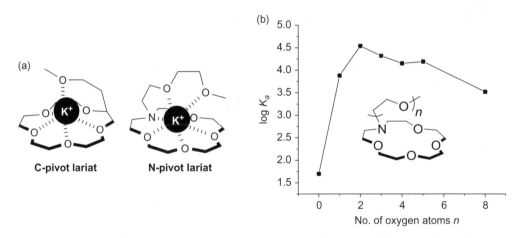

Figure 2.13 (A) Cation complexes of C- and N-pivot lariat ethers; (B) variations in stability constants (in anhydrous methanol) of Na^+ to aza-[15]crown-6 N-pivot lariat ethers with *n* oxygen atoms on the side arm. Figure adapted with permission from R. A. Schultz *et. al., J. Am. Chem. Soc.*, 1985, **107**, 6659. Copyright (1985), American Chemical Society.

Table 2.2 Stability constants of cryptates of various sizes with alkali metal cations in water.

Cryptand	Cavity radius (nm)	LogK				
		Li$^+$	Na$^+$	K$^+$	Rb$^+$	Cs$^+$
[2.1.1]	0.8	5.5	3.2	<2.0	<2.0	<2.0
[2.2.1]	1.1	2.5	5.4	4.0	2.6	<2.0
[2.2.2]	1.4	<2.0	3.9	5.4	4.4	<2.0

Reproduced with permission from: J. M. Lehn & J. P. Sauvage, *J. Am. Chem. Soc.*, 1975, **97**, 6700. Copyright (1975), American Chemical Society.

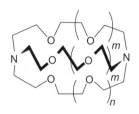

m = 0, *n* = 1: [2.1.1] crytand
m = 1, *n* = 0: [2.2.1] cryptand
m = 1, *n* = 1: [2.2.2] cryptand

Figure 2.14 Nomenclature of cryptands, where each number in the bracket represents the number of oxygen atoms between each nitrogen linker.

Like crown ethers, **cryptands** containing heteroatoms are known. All-nitrogen containing cryptands with either nitrogen or carbon bridgeheads are called *sepulchrates* and *sarcophagines* respectively (Figure 2.15A), and form extremely stable complexes with transition metal cations both thermodynamically and kinetically. As shown in the crystal structure of Co^{3+}-sepulchrate in Figure 2.15B, all six nitrogen atoms on the host coordinates to the Co^{3+} guest in an octahedral stereochemical environment. Other *aza-cryptand* structural variations can exhibit selectivities for main group metals over alkali and transition metal cations. The example (**2.9**) shown in Figure 2.16, reported as recently as 2018 by Paul Beer's group and dubbed the 'Beer can', binds Pb^{2+} more strongly than Na$^+$, K$^+$, Ca^{2+}, and Zn^{2+} in methanol.

Uncomplexed crown ethers and cryptands have a certain degree of conformational flexibility, and their binding sites (Lewis basic atoms) are not convergently oriented. Hence, cation binding will invariably involve some loss of degrees of freedom during ligand conformational rearrangement. Donald Cram realized that to enhance the cation selectivity and affinities even further, highly rigid and preorganized receptors with spatially and geometrically well-defined binding sites were necessary. The outcome was a roughly spherical macrocycle containing octahedrally arranged oxygen atoms oriented towards the interior of the cavity, which he termed a *spherand* (Figure 2.17). Unlike crown ethers and even cryptands, which show some affinities towards less optimally sized cations, spherand **2.10** binds only Li$^+$ and Na$^+$ selectively. Larger alkali metal cations (K$^+$, Rb$^+$ and Cs$^+$) are not bound at all as they are too large to fit within the cavity.

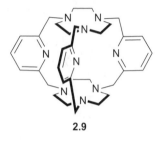

2.9

Figure 2.16 'Beer can' aza-cryptand containing pyridine groups for binding Pb^{2+}.

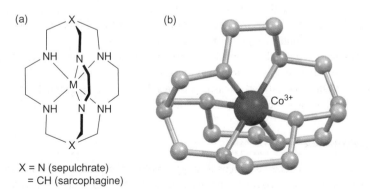

X = N (sepulchrate)
= CH (sarcophagine)

Figure 2.15 (A) Sepulchrates and sarcophagines; (B) crystal structure of a Co^{3+}-sepulchrate complex. Crystal structure refcode: CSD-OAZCOC, first published: I. I. Creaser *et. al.*, *J. Am. Chem. Soc.*, 1977, **99**, 3181.

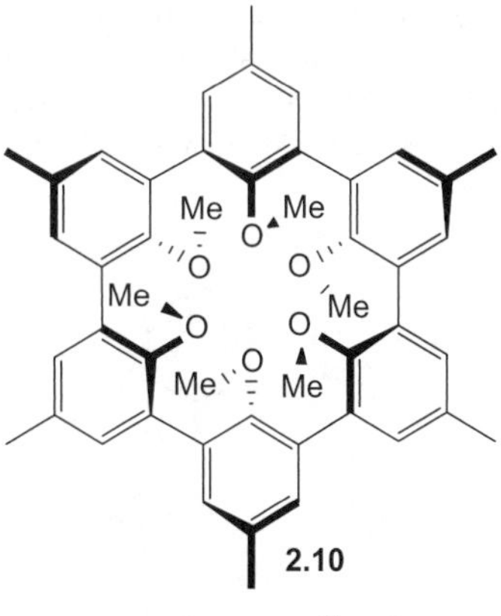

2.10

K_a (Li$^+$) > 7 × 10^{16} M^{-1}
K_a (Na$^+$) = 1.2 × 10^{14} M^{-1}

Figure 2.17 Cram's spherand **2.10**, and its binding constants for Li$^+$ and Na$^+$ (Solvent: D$_2$O-saturated CDCl$_3$). (D. J. Cram & G. M. Lein, *J. Am. Chem. Soc.*, 1985, **107**, 3657.)

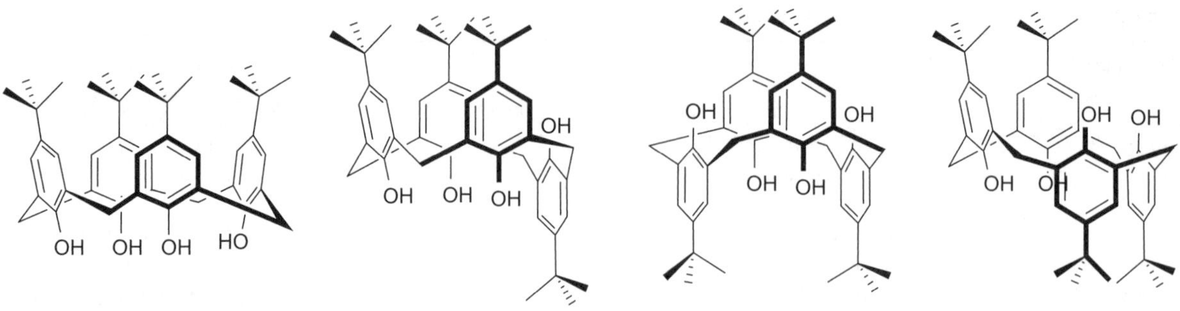

Cone

Partial cone

1,3-alternate

1,2-alternate

2.11

Figure 2.18 Conformations of calix[4]arene.

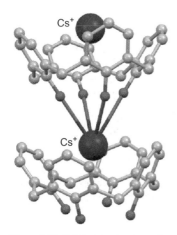

Figure 2.19 Solid-state structure of calix[4]arene with Cs⁺. Crystal structure refcode: CSD-BAMJOP, first published: P. Thuery *et. al., Polyhedron*, 2002, **21**, 2497.

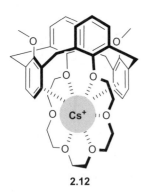

2.12

Figure 2.20 Cs⁺ binding by receptor **2.12** with a calixarene unit in the 1,3-alternate position.

Calixarenes

Calix[n]arenes are a class of synthetic macrocyclic receptors comprising of a cyclic array of *n* phenol units linked by methylene spacers. The name of these molecules is derived from 'calix' (or chalice), owing to their vase-like shapes, and 'arenes' from their aromatic constituents. Calixarenes can adopt numerous conformations. For example, calix[4]arene **2.11** can exhibit four conformations: cone, partial cone, 1,2-alternate, and 1,3-alternate (Figure 2.18), of which only the cone and 1,3-alternate conformations are most frequently encountered in cation binding. The supramolecular interactions responsible for binding the cation are clearly seen in the solid-state structure of unfunctionalized calix[4]arene **2.11** with Cs⁺, shown in Figure 2.19. The bottom half of Cs⁺ is bound symmetrically in the bowl-shaped cavity of **2.11** by cation–π interactions with the electron rich aromatic rings. At the same time, the top half of the cation is bound by ion–dipole interactions with the oxygen atoms on an adjacent molecule of **2.11**.

The phenolic groups on the bottom lower-rim face of calix[4]arenes can be easily appended with further ligating groups to bind numerous cations. For instance, by attaching a crown ether, Cs⁺ was observed to be bound via a 1,3-alternate conformation by receptor **2.12** (Figure 2.20) instead of the cone conformation we have seen in Figure 2.18. Tetraamide functionalization of the lower-rim also allows binding of lanthanide cations such as the neodymium complex of **2.13** seen in Figure 2.21A, where Nd^{3+} is bound by eight convergent oxygen atoms. Two calix[4]arene motifs can also be joined covalently in a lower-rim manner by appropriate linkers to form a cage-like structure. These molecules, dubbed 'calixtubes', display excellent selectivity for alkali metal cations. The calix[4]tube **2.14**, joined by four ethylene spacers, shows exceptional selectivity for K⁺ over other alkali metal cations (Figure 2.21B).

Siderophores

Iron is an essential trace element necessary for biological processes such as respiration and DNA synthesis. However, the natural bioavailability of iron as Fe^{3+} is very low due to the extremely poor aqueous solubility of the cation (~ 10^{-18} M). Hence, plants and bacteria produce ligands known as *siderophores* to scavenge for Fe^{3+} and facilitate uptake across cell membranes by active transport. These structurally diverse, naturally

(a)

(b)

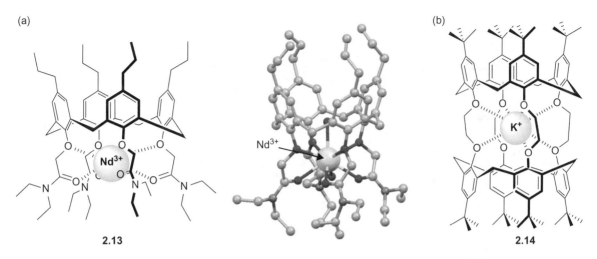

2.13

2.14

Figure 2.21 (A) Chemical and crystal structures of the neodymium complex of tetraamide-functionalized calix[4]arene **2.13**. (Crystal structure refcode: CSD-TAFCIP, first published: C. R. Driscoll *et. al.*, *Supramol. Chem.*, 2016, **28**, 567.) (B) A calix[4]tube **2.14** and its K+ complex.

occurring molecules utilize 'hard' oxygen ligating groups to bind the 'hard' Fe^{3+} cation by forming extremely stable octahedral complexes. For example, the enterobactin siderophore **2.15** produced by *E. coli* bacteria binds Fe^{3+} with three catecholate ligands (Figure 2.22), with an impressive binding constant of 10^{52} M^{-1} in water! Other than Fe^{3+}, some siderophores show affinity towards other 'hard' multivalent cations such as Al^{3+} and Ga^{3+}.

The strong affinities of multivalent cations by siderophores have spurred the development of artificial analogues for biomedical applications. In treatments of diseases such as sickle-cell anaemia, an iron overload in the body can sometimes result. Siderophores have been explored as a means to remove the excessive metal by iron chelation therapy. Most notably, Kenneth Raymond from the University of California, Berkeley, has reported synthetic siderophores capable of complexing Fe^{3+} strongly. Inspired by the natural catecholate Fe^{3+} binding motif, Raymond's siderophore designs similarly allow the molecules to wrap around the cation to form stable octahedral complexes, as illustrated by example **2.16** in Figure 2.23A. A particularly interesting recent application of siderophores has been found for targeted antibiotic treatment. Capitalizing on the fact that bacteria can recognize siderophores as Fe^{3+}-delivery agents much better than mammalian cells, covalently linking an appropriate drug (e.g. antibiotic) onto a siderophore can allow selective uptake of the drug by bacteria cells (**2.17** in Figure 2.23B). This 'Trojan horse' strategy ultimately leads to selective bactericide.

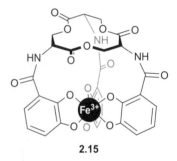

2.15

Figure 2.22 Fe^{3+} complexation by the three catecholate groups of the enterobactin siderophore.

Receptors for binding organic cations

The vast majority of the cation receptors we have considered so far were employed to bind monoatomic metal cations. However, many cationic organic compounds exist which play important functions in biology and nanotechnology. Taking into account the geometric and physical properties of target organic cationic guest species, this section considers the major classes of receptors developed to achieve their strong binding.

Compared with metal cations, organic cations are structurally more complex, hydrophobic, and positive charge distribution is not spherical. Consider the case of the

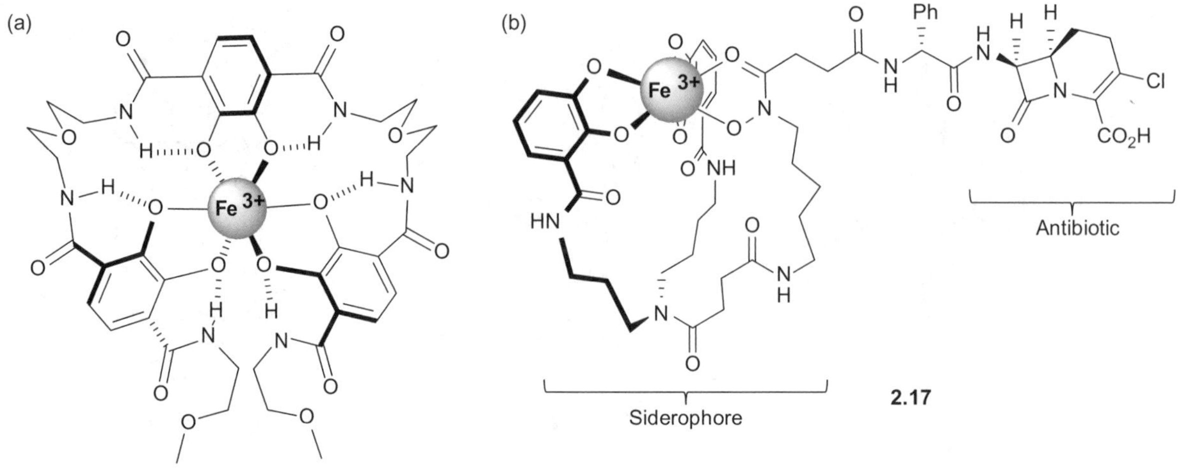

(a)

(b)

Antibiotic

Siderophore

2.17

2.16

Figure 2.23 Biomedical applications of siderophores: (A) Fe^{3+} complexation by a synthetic siderophore for iron chelation therapy; (B) drug–siderophore conjugate for targeted antibiotic therapy.

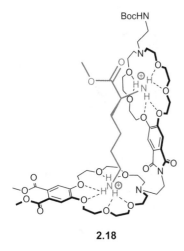

2.18

Figure 2.24 Recognition of dicationic organic compounds using a ditopic receptor **2.18** containing two crown ether motifs.

dicationic lysine methyl ester shown in Figure 2.24, whose positive charges reside on the ammonium moieties on opposite ends of the molecule separated by a C5-alkyl spacer. To bind this dication, a ditopic receptor **2.18** was designed, where crown ether motifs situated on opposite ends of the receptor molecule are at an optimal distance to interact with both ammonium groups. At the same time, the hydrophobic alkyl chain of the dicationic guest is positioned adjacent to the hydrophobic phthalimide motif of the receptor. Impressively, **2.18** binds lysine methyl ester strongly (log$K$ > 4 for each crown ether unit) in highly competitive protic solvents such as methanol and water.

The hydrophobicity of organic cations can be exploited for very strong binding in water using macrocyclic receptors containing deep hydrophobic pockets. Cucurbit[n]urils (abbreviated CB[n]) are a synthetic class of these molecules which are shaped like empty barrels open on both ends, where n denotes the number of glycoluril repeating units (Figure 2.25). Each glycoluril unit is linked by a methylene ($-CH_2-$) spacer to form a rigid structure, and each 'opening' is rimmed by carbonyl oxygens. The internal cavity of CB[n] is unusually hydrophobic due to the lack of any functional groups and lone pairs, whose exceptional low polarizability is more akin to the gas phase than any known organic solvent. Hence organic cations such as quaternary ammonium and imidazolium species can be bound by CB[n] in aqueous solvent media with their hydrophobic units encapsulated within the macrocycle cavity, whilst the charged cationic groups interact with the electron rich oxygen atoms on the 'openings' via ion–dipole interactions. It has also been shown that expulsion of water molecules within the CB[n] cavity provides an important entropic driving force for binding these hydrophobic cations.

Although five, six, seven, eight, and ten-membered CBs are known, CB[7] is arguably the most important member of this family. Compared to other CBs which have generally poor solubility in water (<10^{-5} M), CB[7] exhibits moderate aqueous solubility of ca. 20–30 mM. Furthermore, its size allows it to encapsulate a number of organic cations very strongly, some of which exceed even the strongest host–guest complexes known in nature. Some examples of exceptionally stable complexes in water include

CB[7] with ferrocene derivative **2.19** ($K_a > 10^{15}$ M^{-1}) (Figure 2.26A), and the diamantane quarternary diammonium cation **2.20** ($K_a > 10^{17}$ M^{-1}) (Figure 2.26B). The crystal structure of the latter shows that the diamantane core is complementary to the internal cavity of CB[7], whilst the ammonium groups allow 14 optimal ion–dipole interactions with the carbonyl oxygens.

2.5 Anion binding receptors

Unlike cation binding, the design of receptors to bind anions is a much more recent development. We have already seen how anion binding can be more challenging than cations in the section on differences between cations and anions (in Section 2.3) owing to their comparatively larger size, greater charge diffusivity, range of geometries, and pH dependence. In this section, we shall see how supramolecular chemists have devised ingenious methods to overcome these challenges to selectively bind different anions possessing various physicochemical properties. While the following sections are organized by the dominant supramolecular interaction exploited for anion binding, the reader should bear in mind that combinations of interactions are commonly exploited to anchor the anion tightly within receptor binding sites. For example, electrostatic attractions between cationic receptors and anions often work in synergy with interactions such as hydrogen bonding to strengthen anion binding.

Anion binding by electrostatic interactions

Electrostatic interactions are the simplest means of anion binding, which involves the Coulombic attraction between a positively charged host and a negatively charged guest. Electrostatic attractions are very strong and have a longer range than other non-covalent interactions, but are non-directional. Receptors can be constructed which contain an array of cationic groups, but their mutual electrostatic repulsion can result in design problems. Another important problem arises from the necessity of charge balance: for every cationic group on a receptor, there must be an associated counteranion present. These anions can interact strongly with the receptor and interfere with the binding of the desired anionic guest. Schmidtchen and co-workers solved these problems by synthesizing a **zwitterionic** cage **2.21** containing quarternary ammonium units at each of its four corners (Figure 2.27). The cage structure of

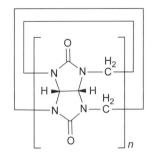

Figure 2.25 Cucurbit[n]urils and their electronic features allowing binding of organic cations.

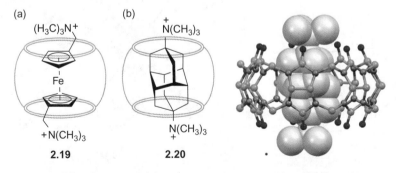

Figure 2.26 CB[7] complexes with (A) diammonium ferrocene **2.19** and (B) diamantane quarternary diammonium cation **2.20** with its crystal structure. (refcode: CSD-FITBIV, first published: L. Cao *et. al., Angew. Chem. Int. Ed.*, 2014, **53**, 988.)

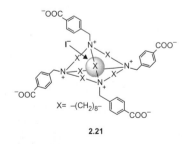

Figure 2.27 Iodide binding in water within Schmidtchen's zwitterionic cage **2.21**.

the host creates a rigid structure which prevents structural distortion of the receptor by electrostatic repulsion, while at the same time creating a confined cavity capable of hosting an anion. Each cationic ammonium vertex is also linked to an anionic benzoate unit, which balances the charges to create an overall neutral zwitterionic compound. This design eliminates the necessity of having a competitive counteranion. Due to these design features, cage **2.21** can bind iodide with impressive strength in water within its cationic cage. The affinity of halides to **2.21** decreases in the order of $I^- > Br^- > Cl^-$ as iodide has the most optimal spatial fit within the cage.

Hydrogen bonding anion receptors

Hydrogen bonding is the most ubiquitous supramolecular interaction exploited for anion binding. Compared to electrostatic interactions, hydrogen bonding is shorter in range (the strength E of ion–dipole interactions $\propto r^{-2}$) but more directional, being strongest when completely linear. Anion receptors are often designed containing arrays of hydrogen bonding motifs, with the hydrogen donor atoms oriented convergently towards the centre of the binding cavity. An enormous diversity of hydrogen bonding motifs and receptors have been developed, and this section can only provide a snapshot of the field (see Section 2.9, Further reading for more in-depth reviews).

The directionality of hydrogen bonding allows the rational design of receptors with complementary shapes to their target anions for enhanced selectivity. We have seen an example of a C_{3v} receptor designed to match the shape of the trigonal-planar NO_3^- anion in Figure 2.6. Receptor **2.4** contains amide linkages, which are amongst the most common hydrogen bonding motifs due to their anion binding potency. Furthermore, they can be easily synthesized from amines and carboxylic acids. The group of Stefan Kubik has exploited amide linkages extensively in their cyclopeptide anion receptors. In the example shown in Figure 2.28, the convergent amide linkages of cyclopeptide **2.22** can bind the large iodide anion to form a 2:1 host–guest complex with the iodide

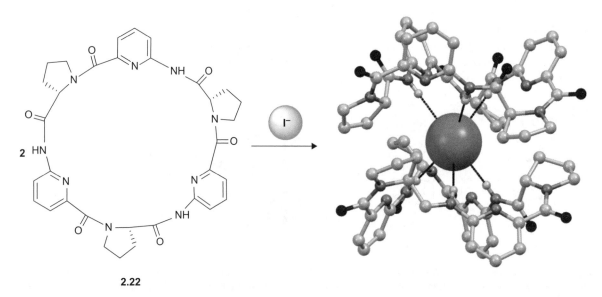

2.22

Figure 2.28 Kubik's cyclopeptide for binding iodide. (Crystal structure refcode: CSD-XITDIN, first published: S. Kubik & R. Goddard, *PNAS*, 2002, **99**, 5127.)

sandwiched between two macrocycles, which effectively shields the anion from the surrounding aqueous solvent.

The hydrogen bonding acidity of the amide motif can be enhanced by replacing the oxygen atom with sulfur, forming a thioamide. This has been attributed to the improved delocalization of the nitrogen lone pair into the C=S bond. Indeed, thioamide receptors have greater affinities for anions than (oxo)amides: the K_a of thioamide macrocycle **2.23S** (Figure 2.29) for $H_2PO_4^-$ in DMSO is more than two orders of magnitude larger than that of the amide analogue **2.23O**.

Urea and thiourea groups are especially effective for binding oxoanions such as carboxylates and phosphates. This is because the orientation of the bidentate hydrogen atoms perfectly complements these oxoanions to form highly linear hydrogen bonds, which maximizes their strength (Figure 2.30A). Like amides/thioamides, thioureas are more potent hydrogen bond donors than ureas (Figure 2.30B), allowing stronger anion binding even in polar solvents such as DMSO.

Squaramides are a class of anion binding motifs gaining increasing importance. These motifs possess a highly rigid four-membered ring with pendant NH units outside the ring which are capable of convergent hydrogen bonding with anions (Figure 2.31A). The extensive delocalization of the nitrogen lone pairs into the conjugated planar four-membered ring causes the NH units to become very potent hydrogen bond donors. Indeed, squaramide receptors are known to bind halides such as Cl^- more strongly than analogous urea and thiourea receptors in competitive DMSO–water mixtures. Very recently, macrocycle **2.26** containing three convergent squaramide units (Figure 2.31B) has been shown to bind SO_4^{2-} selectively in 2:1 water/DMSO solvent. This is especially impressive considering the anion's hydrophilicity.

Macrocyclic receptors containing heterocyclic rings such as pyrroles are another extensive family of N–H hydrogen bonding anion receptors. Pioneered by Jonathan Sessler's group, **Calix[n]pyrroles** are the most important member of this family, comprising of n pyrrole units arranged cyclically within a macrocycle. Like calix[n]arenes, calix[n]pyrroles show conformational flexibility. In the absence of an anion guest, calix[4]pyrrole **2.27** can adopt the 1,3-alternate conformation with consecutive pyrrole units oriented up and down alternately (Figure 2.32). Fluoride and chloride binding, however, forces the receptor to adopt a cone conformation where all pyrrole units form convergent N–H hydrogen bonds to the halide anion, perched above the plane of the ring. Anion affinity can be further increased by introducing a 'strap' containing further hydrogen bonding motifs above the plane of the calix[n]pyrrole ring. Compared with **2.27**, pyrrole-strapped calix[4]pyrrole **2.28** exhibits nearly two orders enhancement in magnitude of Cl^- affinity in acetonitrile (Figure 2.33).

The strength of anion binding by hydrogen bonding receptors can be further augmented by electrostatics conferred by cationic groups on the receptor. Numerous strategies have been designed to achieve this. Firstly, polyaza-receptors can be protonated under acidic conditions to yield cationic ammonium groups which are potent N–H hydrogen bond donors. In a famous example by Lehn, an ellipsoidal hexaprotonated cryptand **2.29** was shown to coordinate the structurally complementary linear azide anion (N_3^-) within its cavity (Figure 2.34A). The azide anion is held by two pyramidal arrays of three hydrogen bonds from the protonated ammonium groups on both ends of the cryptand to each terminal nitrogen atom on the anion. Another approach to using protonated nitrogen atoms for anion binding has been developed by Sessler and co-workers. By expanding the ring size of porphyrins, the basicity of the pyrrolic nitrogen atoms in the larger sapphyrin macrocycles can be increased, allowing

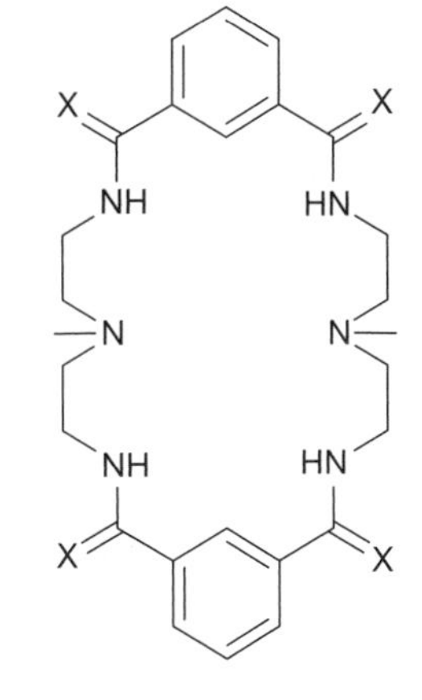

$X = S$ (**2.23S**)
$= O$ (**2.23O**)

Figure 2.29 Thioamide macrocycle **2.23S** and its amide analogue **2.23O**.

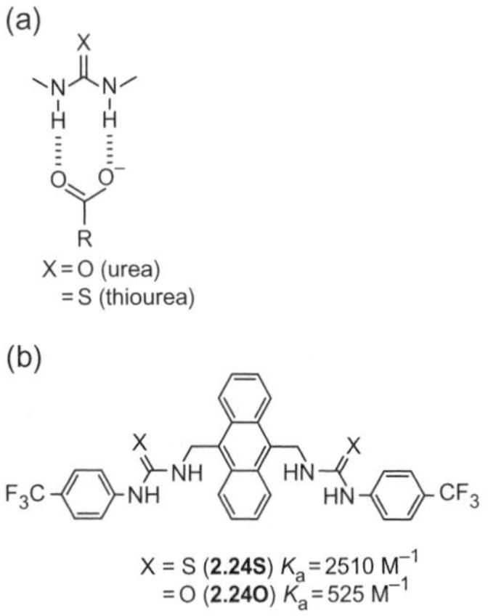

(a)

$X = O$ (urea)
$= S$ (thiourea)

(b)

$X = S$ (**2.24S**) $K_a = 2510$ M^{-1}
$= O$ (**2.24O**) $K_a = 525$ M^{-1}

Figure 2.30 (A) Complementary geometry of binding between the urea/thiourea motif with oxoanions; (B) urea and thiourea receptors for pyrophosphate ($P_2O_7^{4-}$) in DMSO. (T. Gunnlaugsson et. al., *Org. Biomol. Chem.*, 2005, **3**, 48)

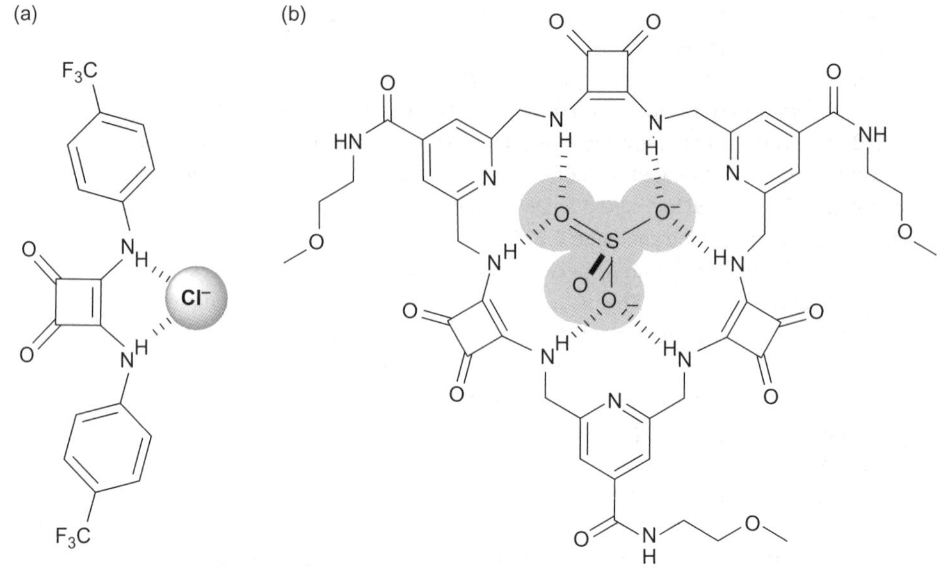

(a)   (b)

2.25   **2.26**

Figure 2.31 (A) A squaramide receptor and its Cl⁻ complex; (B) a squaramide macrocycle selective for SO_4^{2-}.

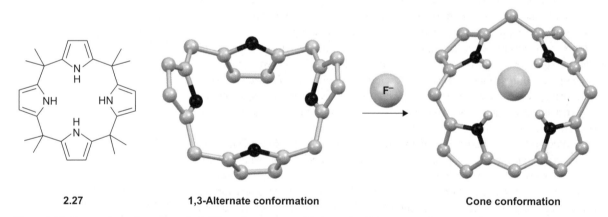

2.27   **1,3-Alternate conformation**   **Cone conformation**

Figure 2.32 Changes in conformation of calix[4]pyrrole upon anion binding. All additional methyl groups and hydrogen atoms removed for clarity. (Crystal structure refcodes: CSD-VUSFIY01 and CSD-TEQKIJ, first published: P. A. Gale et. al., J. Am. Chem. Soc., 1996, **118**, 5140.)

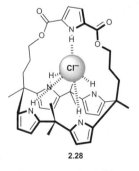

2.28

Figure 2.33 Strapped calix[4]pyrrole.

them to be protonated under acidic conditions to form large electron-deficient binding cavities. Diprotonated sapphyrin **2.30**, for instance, binds F⁻ within the plane of its cavity by five convergent hydrogen bonds (Figure 2.34B).

Despite the effectiveness of these receptors for binding anions, their potency is restricted to low or moderate pH values when the nitrogen atoms remain protonated (pK_a of secondary amines ≈ 10.5). Guanidine groups, on the other hand, are more basic (pK_a ≈ 13.6). Hence, they can remain protonated up to higher pH values, extending the pH range for electrostatics and hydrogen bonding to work hand-in-hand. Like ureas and thioureas, the hydrogen bonding donor structure of guanidiniums make them especially suitable for bidentate binding of oxoanions

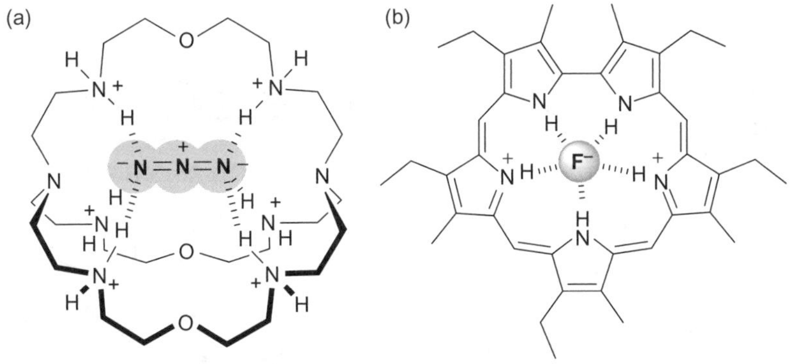

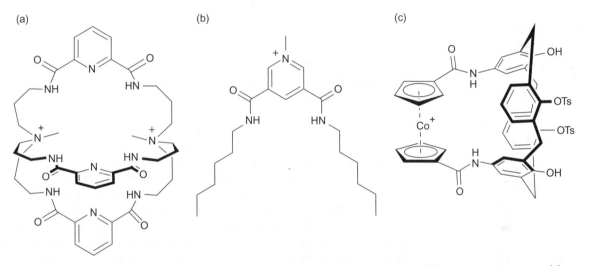

(a)

2.29

(b)

2.30

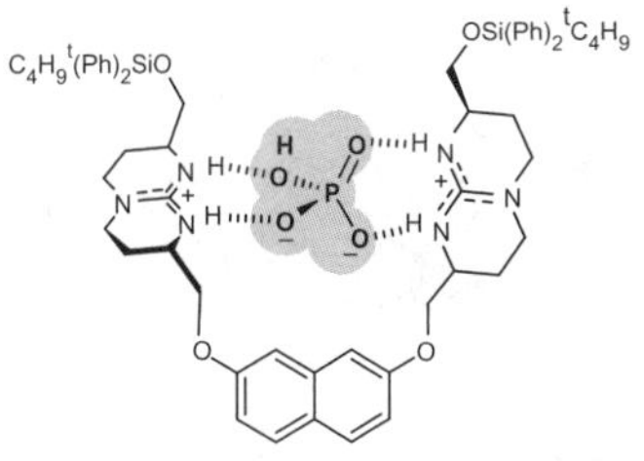

2.31

Figure 2.35 A guanidium receptor for HPO$_4^{2-}$.

Figure 2.34 (A) An octaaza-cryptand for azide binding; (B) a doubly protonated sapphyrin receptor for binding F$^-$.

(a)

(b)

(c)

Figure 2.36 Cationic hydrogen bonding anion receptors with permanently charged cationic units: (A) quarternary ammoniums; (B) pyridinium; (C) cationic metal complexes (e.g. cobaltocene). Counteranions are omitted in this figure.

such as phosphates, sulfates, and carboxylates. The bis-guanidinium receptor **2.31**, shown in Figure 2.35, is capable of binding the biologically-important HPO$_4^{2-}$ and 5′-adenosine monophosphate anions very strongly (K$_a$ > 10^4 M^{-1}) in methanol. Alternatively, to completely remove the possibility of deprotonation, permanent cationic groups can be incorporated into the receptor structure. This can take the form of cationic organic functional groups such as quarternary ammoniums, as demonstrated by Schmidtchen's cage **2.21** in Figure 2.27 and the macrocycle in Figure 2.36A, methylpyridinium (Figure 2.36B), or cationic organometallic species such as cobaltocenium (Figure 2.36C).

C–H hydrogen bond donors are receiving increasing attention for their anion coordination abilities. An example of a recently developed campestarene macrocycle capable of binding Cl$^-$ within its cavity by convergent C–H hydrogen bonds was encountered in Figure 1.8A (Chapter 1). This macrocycle has also been recently demonstrated to stabilize a dimer of bisulfate (hydrogen sulfate) anions (Figure 2.37),

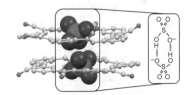

Figure 2.37 A bisulfate anion dimer stabilized by C–H hydrogen bonding campestarene macrocycles (see Figure 1.8A (Chapter 1) for structure). (Crystal structure refcode: CSD-IYEFAV, first published: E. M. Fatila *et. al.*, *Angew. Chem. Int. Ed.*, 2016, **55**, 14057.)

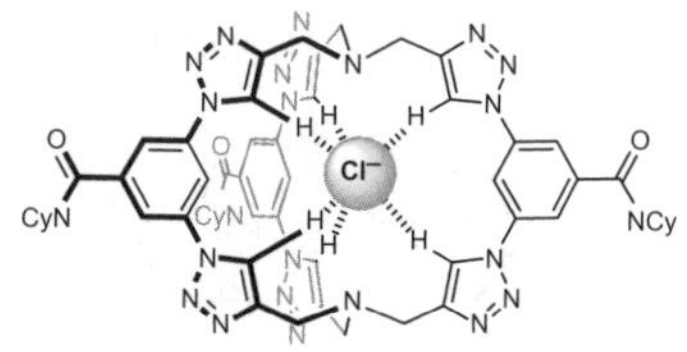

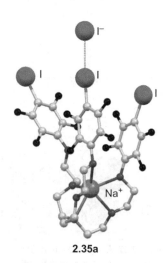

Figure 2.38 A hexa-triazole cryptand capable of extremely strong Cl⁻ binding (Cy = cyclohexyl).

2.32

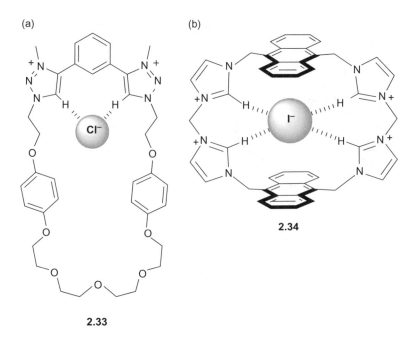

(a)

(b)

2.34

2.33

Figure 2.39. (A) A bis-triazolium macrocycle for Cl⁻ binding and (B) a tetra-imidazolium macrocycle for I⁻ binding.

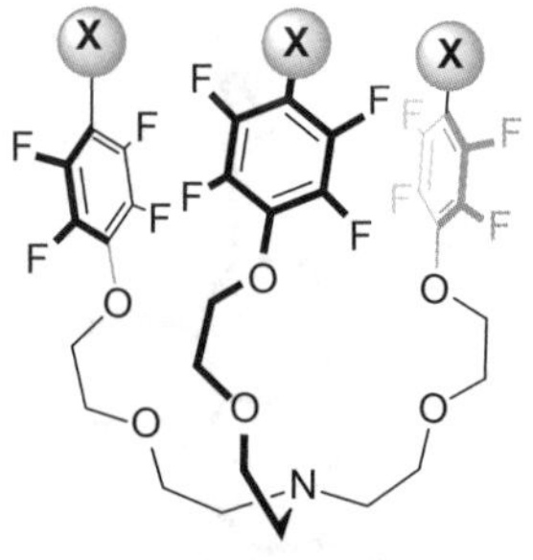

X = I (**2.35a**)
  = F (**2.35b**)

2.35a

Figure 2.40 The first halogen bonding receptor and its monodentate iodide coordination. (Crystal structure refcode CSD-DAXGIT, first published: A. Mele *et. al.*, *J. Am. Chem. Soc.*, 2005, **127**, 14972.)

overcoming the mutual electrostatic repulsion between the anions due to the extensive C–H hydrogen bonding network conferred by the macrocycles.

1,2,3-Triazoles are especially useful C–H hydrogen bond donors due to their electron-deficient nature conferred by the three contiguous nitrogen atoms, as well as their ease of synthesis from azides and alkynes. Incredibly, a cryptand **2.32** containing six triazole units showed a Cl⁻ affinity of 10^{17} M⁻¹, determined by extraction experiments from water into dichloromethane (Figure 2.38). The exceptional affinity and selectivity seen over larger halides and nitrate was due to the extremely rigid and preorganized binding cavity, with the bound Cl⁻ anion held by six convergent C–H hydrogen bonds from the triazoles. Another very useful feature of triazoles is their capacity to be alkylated at the N^3 position to form triazoliums with even more polarized C–H bonds. Receptors containing triazoliums can be potent hydrogen bonding anion receptors, such as the bis-triazolium macrocycle **2.33** selective for Cl⁻ in acetonitrile (Figure 2.39A). Finally, like triazoliums, imidazoliums are five-membered aromatic heterocycles possessing a highly polarized C–H bond capable of interacting strongly with anions. This property has been exploited to synthesize a cyclophane with four imidazolium groups (Figure 2.39B) which binds I⁻ with $K_a > 10^4$ M⁻¹ very strongly in aqueous solution at pH 7.4. These C–H hydrogen bond donors have the notable advantage of being resistant to deprotonation, enabling them to retain their anion binding potency in a wide range of pH.

Halogen bonding anion receptors

In 2005, Metrangolo and Resnati first demonstrated the potential of halogen bonds for anion recognition with receptor **2.35a** (Figure 2.40), containing three iodoperfluorobenzene units. From NMR experiments in CDCl₃, they determined that **2.35a** could complex NaI with much higher affinity than the all-fluorinated analogue **2.35b**. In the solid state, they observed a highly linear

(176°) C–I···I⁻ halogen bond from one of **2.35a** receptor's iodoperfluorobenzene units (Figure 2.40). Since this seminal work, the last decade has witnessed an explosive development of halogen bonding receptors. Uniquely, halogen bonding allows the precise control of anion binding geometry due to the stringent linear directional nature of the interaction, much more so than hydrogen bonding. While the length of this primer prevents an in-depth discussion of this fascinating field (see Section 2.9, Further reading for more detailed discussion), halogen bonding receptors also show numerous advantages over hydrogen bonding analogues. Frequently, halogen bonding receptors (with iodine as the donor atom) show higher binding affinities, as well as a greater preference for binding 'softer' and more polarizable anions such as I⁻, especially in aqueous media. While most of the halogen bonding motifs studied use iodine atoms as the donor due to its greater polarizability, bromine is also occasionally employed.

Substitution of the H atom of the electron-deficient triazole motif results in potent iodotriazole (5-iodo-1,2,3-triazole) halogen bond donors for anion binding, often found to be more potent than hydrogen bonding triazole analogues. With the picket fence-type receptor **2.36** (Figure 2.41A) for instance, the halogen bonding receptor shows almost an order-of-magnitude greater K_a for Cl⁻ than the hydrogen bonding analogue in acetonitrile, with smaller but significant enhancements for Br⁻ (3.0 ×) and I⁻ (1.7 ×). The iodotriazole motif is also especially useful for receptor preorganization by intramolecular hydrogen bonding (Figure 2.41B) and metal coordination (Figure 2.41C) to enhance anion binding affinities. For **2.38**, the coordination of rhenium(I) also serves to further polarize the C–I bonds to strengthen halogen bonding interactions.

Likewise for hydrogen bonding receptors, the addition of cationic motifs results in *charge-assisted* halogen bonding receptors capable of binding anions in competitive aqueous solvent media. Iodotriazoliums, formed by N-methylation of the triazole nitrogen atom (**2.39** in Figure 2.42A), can form halogen bonding motifs potent enough to bind halides (K_a of I⁻ > Br⁻ > Cl⁻) in pure water. Haloimidazoliums are also highly effective, with macrocycle **2.40** (Figure 2.42B), possessing two bromoimidazolium units, exhibiting selectivity for Br⁻ in a solvent mixture containing 10% water in methanol.

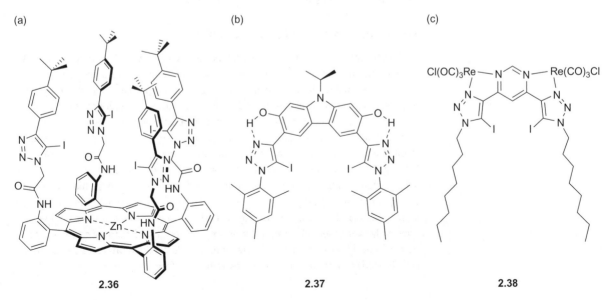

(a)   (b)   (c)

2.36   2.37   2.38

Figure 2.41 Iodotriazole-containing halogen bonding receptors for anion binding.

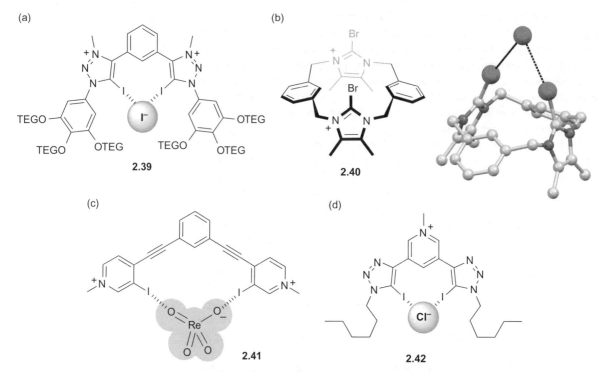

(a)

TEGO

TEGO   OTEG   TEGO   OTEG

OTEG

2.39

(b)

2.40

(c)

2.41

(d)

Cl⁻

2.42

Figure 2.42 Anion binding receptors containing charge-assisted halogen bonding motifs: (A) iodotriazoliums (TEG = tetraethylene-glycol); (B) bromo-imidazoliums and a crystal structure showing a bound bromide anion (refcode CSD-ALOKUI, first published: A. Caballero *et. al., Angew. Chem. Int. Ed.*, 2011, **50**, 1845); (C) and (D) pyridinium-containing donors.

Finally, covalently attaching halogen bonding motifs such as iodine atoms (Figure 2.42C) and iodotriazoles to cationic pyridinium groups (Figure 2.42D) have also been shown to bind anions strongly in competitive solvent media.

Receptors containing Lewis acidic sites

Binding highly hydrated anions such as fluoride, phosphates, and carboxylates in water necessitates the use of very strong interactions to outcompete the extensive anion hydration. Arguably, direct coordination to a 'hard' Lewis acidic metal centre is the strongest interaction available to bind these anions in water, although whether these interactions can be considered strictly supramolecular is debatable. In the receptors, these metal centres are bound by organic ligands with high kinetic and thermodynamic stability, with at least one vacant or weakly bound coordination site available for anion binding. The use of trivalent lanthanide metal centres for anion binding has received a lot of attention lately, particularly for their usefulness to also sense these anions by changes in luminescence intensity (see the section on ion sensing in Section 2.7 later in this chapter). The europium complex **2.43** (Figure 2.43A) binds fluoride strongly and selectively over other halides in water, as only the Ln–F bond is strong enough to compete effectively with the Ln–OH$_2$ interaction. A range of oxoanions can also be strongly coordinated by lanthanide metal centres as they can form stable four- to six-membered

(a)

(b)

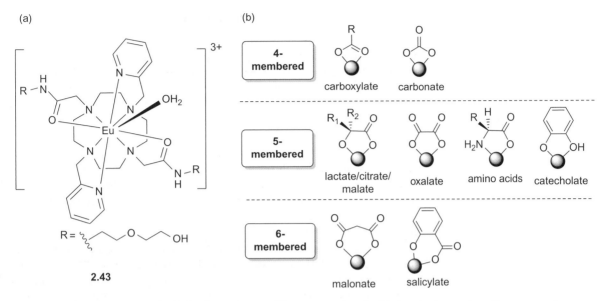

2.43

4-membered	carboxylate	carbonate
5-membered	lactate/citrate/malate, oxalate, amino acids, catecholate
6-membered	malonate, salicylate

Figure 2.43 (A) A europium complex for binding F^- in water; (B) some examples of stable four- to six-membered ternary oxoanion adducts at lanthanide metal centres.

chelate rings (Figure 2.43B). Lewis acidic metal centres, in combination with hydrogen bonding motifs, also present very potent anion binding sites. A Pt^{2+} metal centre chelated with four urea-containing ligands (**2.44**) binds SO_4^{2-} strongly in DMSO via electrostatics conferred by the Pt^{2+} centre and convergent hydrogen bonds from the ureas (Figure 2.44).

The Lewis acidity of boron stems from its empty p-orbital, which readily accepts electron donation from Lewis bases such as anions. Boranes have been heavily exploited to bind nucleophilic anions (e.g. fluoride, azide, cyanide) in competitive protic solvents. Anion binding affinities can be further enhanced by the presence of other Lewis acidic units on the same receptor. Gabbaï and co-workers have designed bidentate anion receptors containing a borane unit and cationic Lewis acidic group, which can chelate the anion of interest between them. Despite their apparent structural simplicity, substituted cationic phosphoniums, stiboniums (containing antimony), and telluroniums (Figure 2.45) can selectively bind fluoride in methanol.

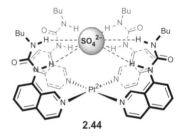

2.44

Figure 2.44 Pt^{2+}-containing receptor for SO_4^{2-}.

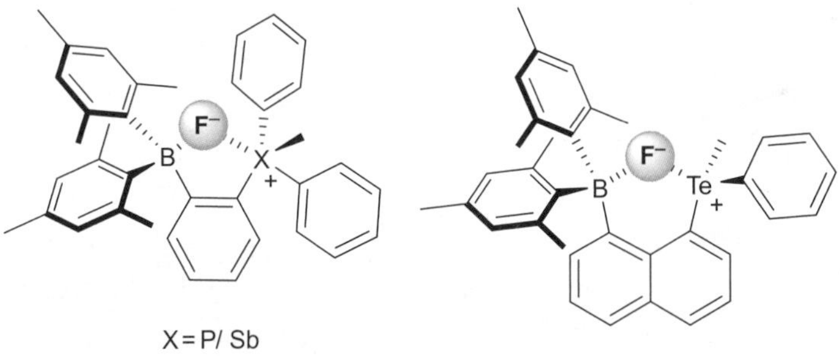

X = P/ Sb

Figure 2.45 Anion receptors containing Lewis acidic boranes.

Anion-π interactions for anion recognition

Anion-π interactions are considerably weaker than most of the supramolecular in-teractions we have surveyed so far, but they can still give rise to appreciable binding affinities when used in combination. This interaction is relatively new in the canon of anion receptor chemistry, with the first crystal structure showing anion-π interactions only reported in 2004. Hence, comparatively few receptors have been designed to exploit it exclusively. Perhaps misleading, the π electrons of the electron-deficient ar-omatic systems do not actually interact with the anion, as this interaction arises mainly from electrostatics and anion-induced polarization effects. Anion-π interactions are strongest in aprotic solvents using small, charge-dense anion guests (F⁻, Cl⁻, OH⁻), with highly electron-deficient aromatic systems such as triazines, perfluoroarenes, and naphthalene diimide (NDI) motifs. In a recent example, macrocycle **2.45** (Figure 2.46A), containing alternate triazine and perfluoroarene groups, binds F⁻ and N_3^- se-lectively within the macrocycle cavity with K_a > 1000 M⁻¹ in acetonitrile. There has been evidence that anion-π interactions can facilitate the transport of anions across membranes as well. Using a series of NDI derivatives containing different degrees of π-acidity and binding site steric hindrance, the effectiveness of Cl⁻ transport cor-related with anion-π interaction strength. Notably, receptor **2.46** (Figure 2.46B) was amongst the most effective in transporting Cl⁻ even at very low concentrations due to its sterically accessible interaction site and highly electron-deficient aromatic system conferred by double cyano-substitution.

Deep-cavity receptors for hydrophobic anions in water

Binding large, charge-diffuse, and hydrophobic anions selectively in water requires a different strategy from what has been considered so far. Consider perchlorate (ClO_4^-), for instance, whose charge-diffusive and poor coordinating nature allows it to be use-ful as a non-coordinating counteranion for metal complexes. For receptors relying on Lewis acidic sites, or hydrogen or halogen bonding, greater affinity will be expected for

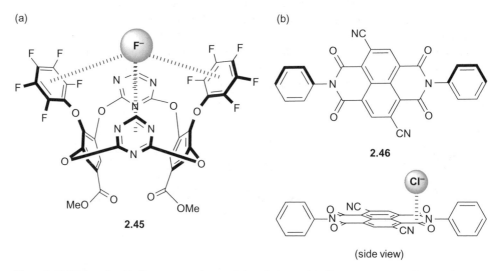

Figure 2.46 (A) An anion binding macrocycle containing triazine and perfluoroarene groups (S. Li *et. al.*, *J. Org. Chem.*, 2012, **77**, 1860); (B) NDI-containing receptor for anion transport.

smaller, charge-dense anions such as Cl⁻ over ClO_4^-. Fortunately, the hydrophobic effect can be exploited to great advantage for binding these large, hydrophobic anions in water. Such receptors typically possess hydrophilic rims to ensure their water solubility, and deep hydrophobic cavities which are shielded structurally from the aqueous environment. These cavities have an inherent preference for easily desolvated hydrophobic anions, in a similar way that CB[7] can bind hydrophobic organic cations very strongly.

The exploitation of the hydrophobic effect for anion binding in water, while presently under-developed, has led to notable successes in recent years. Cavitand **2.47**, possessing a deep hydrophobic pocket (Figure 2.47A), binds ClO_4^- more strongly than even I⁻ in aqueous buffer at pH 11. Hexafluorophosphate PF_6^-, with its even

(a)

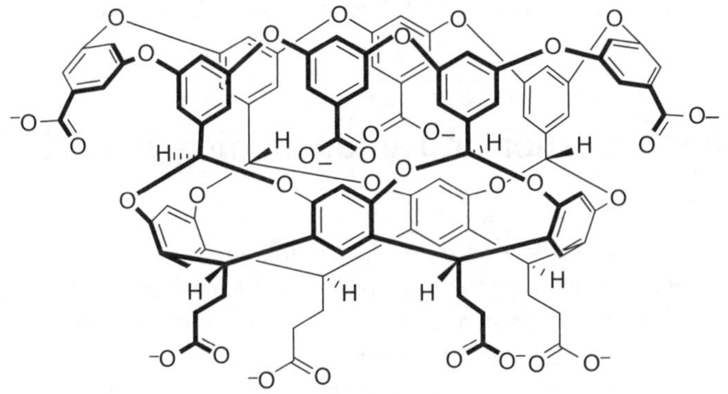

2.47

(b)

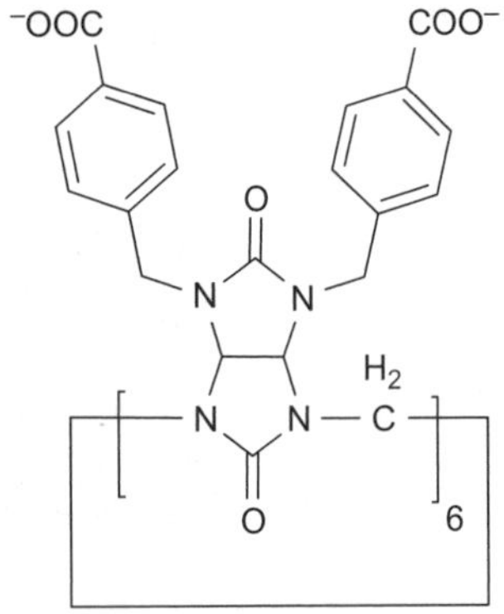

2.48

(c)

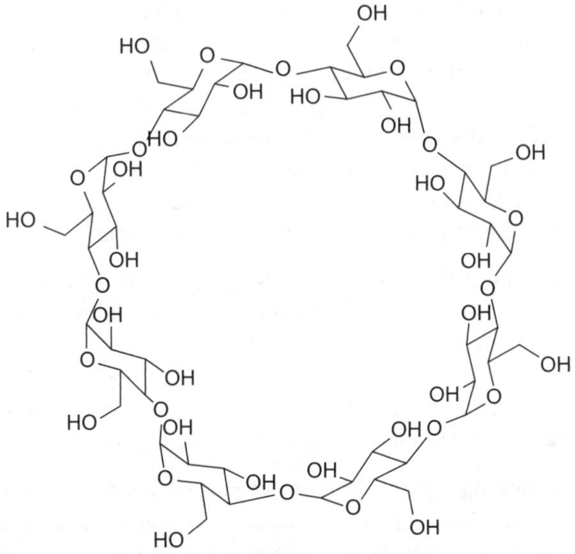

2.49

Figure 2.47 Macrocycles with deep hydrophobic cavities for binding hydrophobic anions in water: (A) cavitand; (B) bambusuril; and (C) γ-cyclodextrin.

greater hydrophobicity than ClO_4^-, is bound even more strongly. Preference for binding hydrophobic anions is observed even in the presence of large excess of hydrophilic salts (e.g. NaF or NaCl). Counter-intuitively, binding affinities for ClO_4^- were observed to be augmented with these salts present. This was attributed to an increase in the ionic strength of the solution by the presence of the co-salts, enhancing the hydrophobic interactions driving ClO_4^- encapsulation. Other types of deep-cavity receptors also show a marked preference for binding hydrophobic anions. Bambusurils are a class of macrocyclic receptors containing substituted glycoluril units (Figure 2.47B) which were first reported in 2010, and are named after their shape resemblance to a bamboo stem. In aqueous buffer at pH 7.1, bambusuril **2.48** exhibits impressive affinities exceeding $10^7 \, M^{-1}$ for I^- and ClO_4^-. A naturally occurring macrocycle containing eight contiguous glucose units, γ-cyclodextrin (**2.49** in Figure 2.47), was shown to bind anionic dodecaborate clusters ($B_{12}Br_{12}^{2-}$) strongly in pure water.

2.6 Simultaneous cation and anion binding (ion pair recognition)

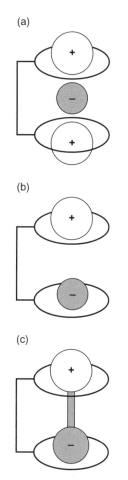

(a)

(b)

(c)

Figure 2.48 (A) Cascade complex; (B) heteroditopic ion-pair binding; and (C) zwitterion binding.

In the previous two sections, we have taken a whirlwind tour of the vast and rapidly growing fields of supramolecular cation and anion binding. We have also seen how creatively exploiting the inherent properties of the ions can allow rational receptor design to achieve good selectivity using the entire toolkit of supramolecular interactions. Notwithstanding, it is important to note that unless the solvent is highly polar and solvating, cations and anions do not exist separately, but invariably as ion pairs in solution. For practical applications such as membrane transport, the separate binding and carriage of either the cation or anion of the ion pair also helps to transport the associated oppositely charged counterion at the same time. An alternative strategy for binding charged guests is to bind *both* the cation and anion together using the same receptor—a concept known as **heteroditopic** *ion pair* recognition. Often, the resulting complex is charge neutral, and this can help with extraction, solubilization, and sensing applications. Furthermore, as we shall soon see, binding of both cations and anions simultaneously can also increase their respective binding affinities compared to single-ion receptors, aided by their mutual electrostatic attraction to each other.

Three fundamentally different approaches to binding ion pairs are known:

The cascade approach (Figure 2.48A): A receptor is designed to bind more than one cation at spatially separate well-defined sites. The bound cations can then bind an anion in the binding cavity between them. There is almost always direct inner-sphere contact between the bound cations and anion to allow *contact ion-pair binding*. Cascade complexes are found in natural metallo-enzymes such as bacterial alkaline phosphatase, where an anionic substrate is bound between two Zn^{2+} centres.

Heteroditopic ion-pair binding (Figure 2.48B): The cation and anion are bound on separate sites of the receptor. Depending on the distance between the cation and anion recognition sites in the receptor design, both ions can be bound as contact ion-pairs, *solvent-bridged ion pairs* (cation and anion are linked by solvent molecules), or receptor-separated ion-pairs (both ions are completely separated from each other with no contact between them). Heteroditopic ion-pair binding accounts for the vast majority of ion-pair receptors available.

Zwitterion binding (Figure 2.48C): Organic molecules, such as amino acids, can possess both acidic and basic groups. At its isoelectric point, the molecule is overall electrically neutral, containing the same number of positively and negatively charged groups, for example ammonium and carboxylate motifs with the amino acid tryptophan (Figure 2.54). Therefore zwitterion recognition necessitates that the receptor design positions cation- and anion-recognition groups in the correct spatial and geometric arrangement to interact with these motifs on the guest molecule.

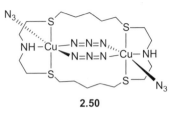

2.50

Figure 2.49 Ellipsoidal cascade complex for binding linear N_3^- anion in water.

Cascade complexes

The importance of cascade complexes to bind strongly hydrated anions in competitive aqueous media was realized during the formative years of supramolecular chemistry. In an early example, the bimetallic Cu^{2+}complexed ellipsoidal macrocyclic ligand **2.50** (Figure 2.49) facilitated the binding of linear anions such as N_3^- to be bridged between both metal centres. Smaller chloride anions, which are unable to bridge the gap between the metal centres, simply coordinate individually to each Cu^{2+}. Azide coordination can occur even in water to form dark green crystals of the complex, as the strong metal–anion interactions can overcome the hydration shell of the anions.

Cascade complexes have become very popular in recent years due to their ability to bind highly basic and hydrophilic oxoanions in water. For instance, zinc(II)-dipicolylamine (Zn(II)DPA) is a widely used motif for binding phosphate species in water due to the strong 'hard–hard' interactions between Zn^{2+} and the anionic phosphate oxygen atoms. Selectivity for the type of phosphate species can be rationally engineered by changing the length of the spacer motifs between the (Zn(II)DPA) motifs. For instance, receptor **2.51** (Figure 2.50A) binds the pyrophosphate anion selectively over ATP^{4-}, ADP^{3-} and SO_4^{2-} in buffered water at pH 7.4. Other than phosphates, carboxylate anions can also be effectively bound by cascade complexes. Acetate, for

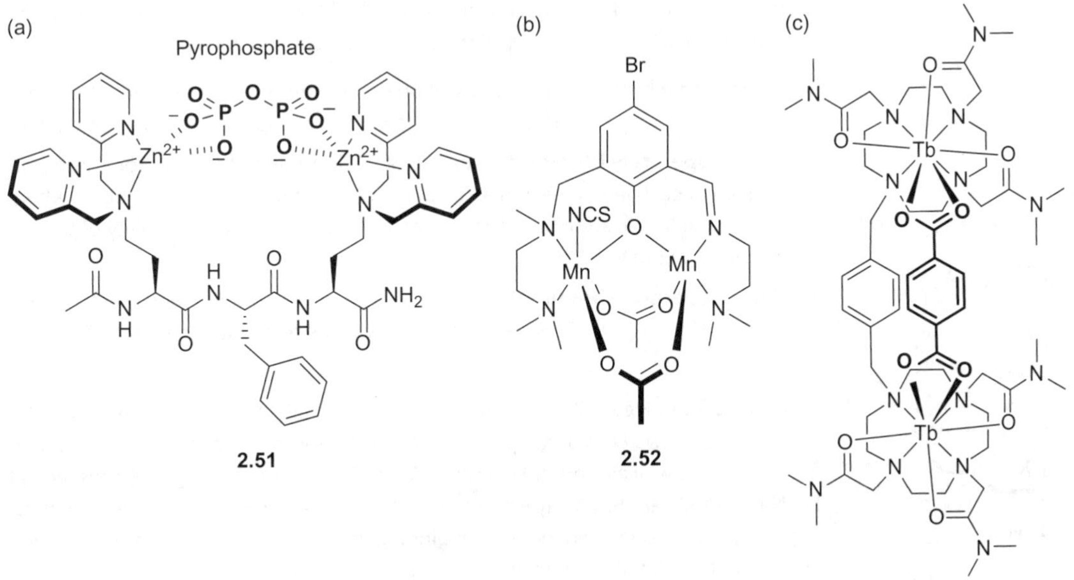

(a)

Pyrophosphate

2.51

(b)

2.52

(c)

2.53

Figure 2.50 Cascade complexes for binding (A) pyrophosphate; (B) acetate; and (C) terephthalate oxoanions.

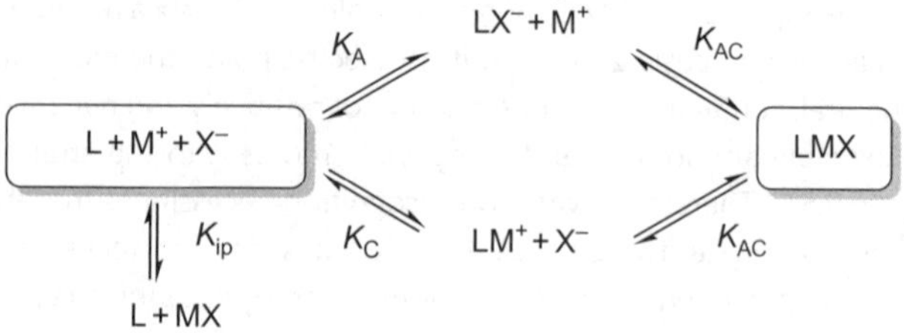

$$L + M^+ + X^- \quad \xrightarrow{K_A} \quad LX^- + M^+ \quad \xrightarrow{K_{AC}} \quad LMX$$

Scheme 2.2 The numerous equilibria involved during ion-pair binding. L = ligand, M^+ = cation, and X^- = anion.

instance, bridges two closely spaced metal centres, with each oxygen atom of the carboxylate group coordinating to one Mn metal cation (e.g. receptor **2.52** in Figure 2.50B). Dicarboxylates, on the other hand, can be complexed using a cascade approach with the two metal centres spaced appropriately further apart. For example, in water complex **2.53** containing two terbium metal centres (Figure 2.50C) binds a terephthalate anion.

Heteroditopic receptors

The binding of ion pairs by heteroditopic receptors is considerably more complicated than binding just an anion or a cation due to the multiple equilibria present. As shown in Scheme 2.2, equilibria exist for the binding of the receptor to anion alone (binding constant denoted by K_A), cation alone (K_C), the association of *both* cation and anion to the receptor (K_{AC}), as well as ion-pair association (K_{ip}) outside the receptor. It should be noted here that ion-pairing outside the receptor is often neglected, and can change the binding affinities of the ion-pair to the receptor drastically, even causing the salt to precipitate from solution should the solvent not be solvating enough! The term, cooperativity factor, is often used to denote how much enhancement in cation/anion binding is observed in the presence of the other bound ion, defined as K_{AC}/K_A or K_{AC}/K_C. The cation and anion can influence each other's binding affinities to the receptor in three possible ways:

- **Cooperative binding**: the binding of one of the ions increases the affinity for the other, i.e. $K_{AC} > K_A + K_C$. This is the most common scenario for ion-pair binding.

- **Anti-cooperative binding**: the affinity of the ion-pair to the receptor is less than the sum of each individual ion association, i.e. $K_{AC} < K_A + K_C$. This can occur when the ion-pair association outside the receptor is stronger than the ability of the receptor to bind to them.

- **Non-cooperative binding**: both cation and anion do not influence each other at all, i.e $K_{AC} = K_A + K_C$. This scenario is very seldom encountered for ion-pair binding.

Receptor **2.54** in Figure 2.51 illustrates the dramatic effects of cooperativity. The receptor contains two N–H donor groups from an indolocarbazole unit capable of anion interactions, as well as a diaza-[18]crown-6 motif for cation binding. The presence of Li^+, Na^+, and K^+ can drastically increase the affinities of halide anions within the binding cavity—up to 2000-fold increase in binding constant value for the case of Cl^- with Na^+. Receptor **2.54** binds the salts as contact ion-pairs, which often leads to very large cooperativity factors. This is due to the favourable ion-pair recognition minimizing the

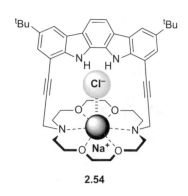

2.54

Figure 2.51 Contact ion-pair binding by receptor **2.54**.

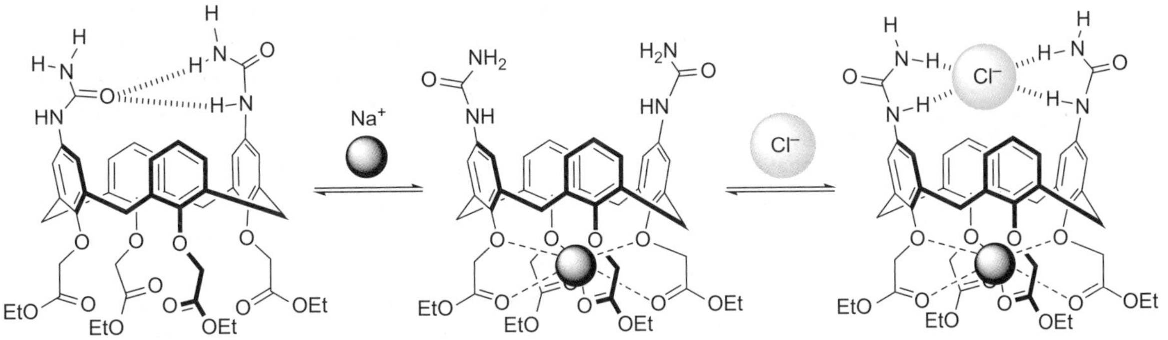

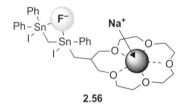

Figure 2.52 Allosteric 'switching on' of halide binding by Na⁺ coordination for receptor-separated ion-pair binding.

2.55

energetic penalty cost of separating the electrostatically bound cation and the anion from each other when binding to the receptor.

Reinhoudt's calix[4]arene-based ion-pair receptor **2.55** (Figure 2.52) is a classic receptor-separated ion-pair receptor, which achieves cation–anion cooperativity very elegantly. The receptor contains anion-binding urea linkages on its upper rim, which in chloroform solution are hydrogen-bonded together intramolecularly, and hence not available to hydrogen bond with potential anion guest species. Sodium cation binding to the ester oxygen atoms on the lower rim of the calix[4]arene, causes lower-rim contraction which results in the urea linkages no longer able to hydrogen bond to themselves in this conformation. This 'opens' up the ureas' anion binding site for strong hydrogen bonding interactions with halides such as Cl⁻ and Br⁻. This overall process of sodium–halide ion-pair complexation is an example of **allosteric binding**, where binding of one species changes the receptor structure so that it affects the binding affinity at a separate site towards a different guest species. Allosteric effects are very common in biology as a form of feedback to augment or diminish the rates of biochemical reactions.

Considering the many useful applications of ion-pair recognition, the field has now expanded to include interactions other than hydrogen bonding. Receptor **2.35a** (Figure 2.40) is an example of ion-pair binding by a halogen bonding heteroditopic receptor. To overcome the high lattice energy of ionic sodium fluoride, an ion-pair receptor containing a Lewis acidic organotin anion binding site has been developed in 2007 (Figure 2.53). With Na⁺ strongly bound within the [15]crown-5 moiety and fluoride coordinated between the tin atoms, sodium fluoride could be solubilized in aprotic polar solvents such as acetonitrile.

Receptors for zwitterions

Zwitterions play important biological roles as naturally occurring amino acids and peptides. At the isoelectronic point of these molecules, they possess equal numbers of positively and negatively charged groups, making them electroneutral. Binding such species under excessively acidic or basic conditions will change their protonation state and transform them into net cationic or anionic species respectively. Receptor **2.57** in Figure 2.54 is a good example of how selectivity for binding an amino acid zwitterion can be engineered. The guanidinium and crown ether motifs of **2.57** bind

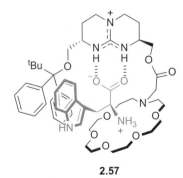

2.56

Figure 2.53 Receptor-separated NaF binding by an organotin receptor.

2.57

Figure 2.54 Chiral-selective three-point interaction between **2.57** and the zwitterionic amino acid tryptophan.

the charged carboxylate and ammonium functional groups respectively, whilst the aromatic groups on the tert-butyldiphenylsilyl (TBDPS) moiety provides selectivity for the R-group of the amino acid. In this case, amino acids containing aromatic side groups such as phenylalanine and tryptophan interact preferentially with the TBDPS group by aromatic π–π stacking interactions, and are more strongly bound than aliphatic amino acids such as glycine and alanine. It is important to note that the three-point host–guest interaction between the amino acid and **2.57** also gives rise to chiral selectivity. The *S,S*-configuration of **2.57** results in a marked preference for binding the L-forms of tryptophan and phenylalanine over their D-forms, due to better geometric and configurational complementary with the chiral receptor.

The presence of *both* anionic and cationic binding sites on the same receptor can sometimes lead to issues such as receptor dimerization, which reduces their capacity to bind to guest molecules. This is especially important if the recognition sites themselves are charged. For instance, to bind a zwitterionic tripeptide, which is important for cell–cell adhesion processes, receptors **2.58** and **2.59** have been designed with the necessary complementary sites for binding. As shown in Figure 2.55A, the flexibility of **2.58**'s structure allows it to adopt a favourable conformation for dimerization, where each receptor binds to each other in an anti-parallel fashion driven by the hydrogen bonding between the bisphosphonate group and the guanidinium motif. On the other hand, substitution of the methylene spacer of **2.58** with benzene in **2.59** makes the receptor's conformation too rigid for dimerization. **2.59** was able to interact with the zwitterionic tripeptide very strongly in buffered water (Figure 2.55B).

2.7 Applications of cation and anion binding

The science of supramolecular host–guest recognition chemistry is ultimately driven by practical applications. In our discussion so far, we have seen how crown ethers can be used as phase-transfer catalysts (Section 2.4: From crown ethers to spherands: increasing selectivity and affinity), siderophores for metal chelation therapy and targeted antibiotic treatment (Section 2.4: Siderophores), and transmembrane transport by anion-π receptors (Section 2.5: Anion-π interactions for anion recognition). Here, we shall survey some of the more important applications of cation and anion binding in greater detail.

Ion sensing

Receptors for charged guests may be designed with *reporter groups*, which are functionalities that give a detectable signal (e.g. electrochemical/redox or optical) when an ion binds to the host. These signals indicate or *sense* the presence of the ions of interest, and in some cases can also be used to monitor their concentrations. Ion sensors can be grouped into three categories depending on their mechanism of action:

1. Receptor-reporter approach (Figure 2.56A): ion sensing occurs through non-covalent host–guest interactions which electronically perturbs the redox/optical properties of the reporter group. These groups are hence designed to be in close proximity to the ion binding site on the host molecule.

2. Indicator displacement assay (IDA) approach (Figure 2.56B): ion binding displaces a dye originally weakly bound to the receptor, leading to a 'turn on/off' optical response.

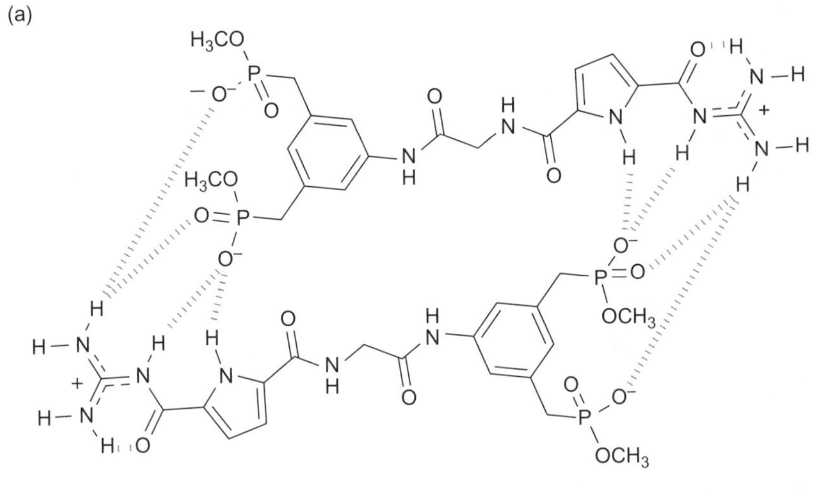

(a)

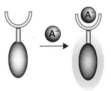

(a) Receptor-reporter sensing

2.58

(b)

(b) Indicator displacement assay

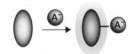

dye

Change in
response from
displaced dye

(c) Chemodosimeter

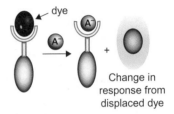

Figure 2.56 The (A) receptor-reporter; (B) indicator displacement assay; and (C) chemodosimeter approaches to ion sensing.

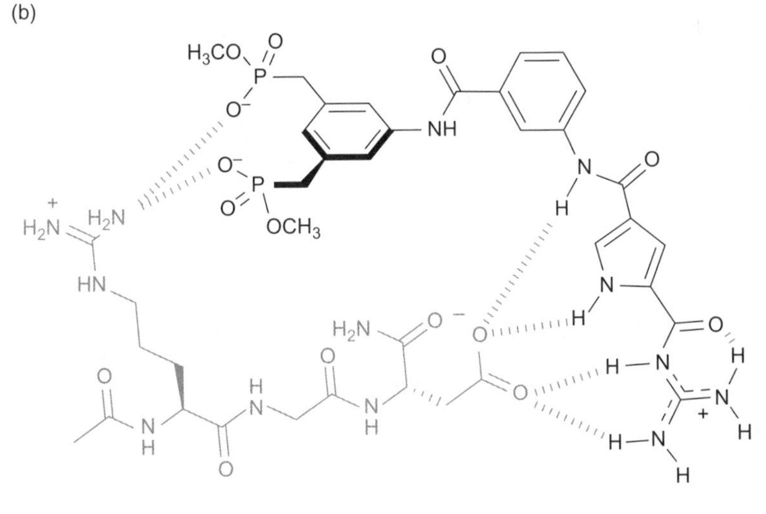

2.59

Figure 2.55 Receptor design for binding tripeptide zwitterions: (A) dimerization of receptor **2.58** with a central methylene spacer; (B) interaction between rigid receptor **2.59** and tripeptide Ac–Arg–Gly–Asp–NH₂.

3. Chemodosimeter approach (Figure 2.56C): covalent bond formation between the anion and receptor, changes the optical and electrochemical properties of the latter.

Receptor-reporter ion sensors are the most common of the three classes. For instance, receptor **2.60** (Figure 2.57A) contains a crown thioether macrocycle which is tethered directly onto a fluorescein motif. Due to favourable 'soft–soft' interactions, **2.60** binds the soft Hg²⁺ cation in water selectively over Cu²⁺, Pb²⁺, and alkali metal cations. Concurrently, Hg²⁺ binding resulted in 170-fold fluorescence enhancement of **2.60.** While fluorescence ion sensing is useful, the ability to detect ions by colour changes

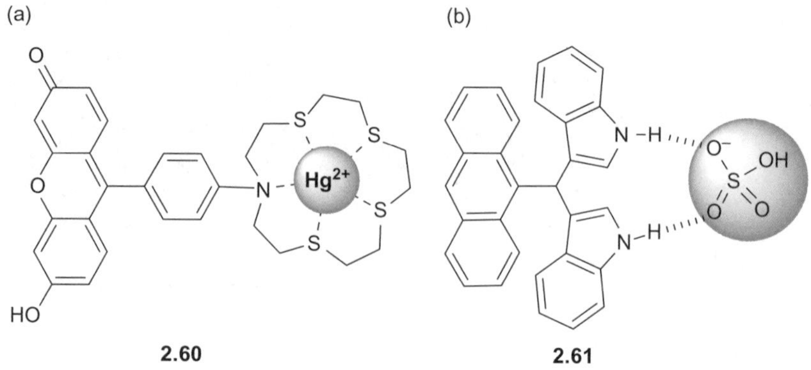

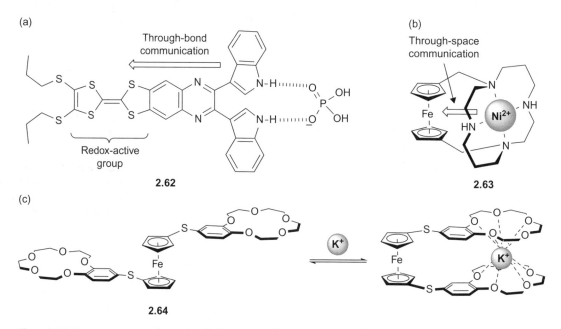

(a)   (b)

2.60   **2.61**

Figure 2.57 Fluorescent optical sensors for (A) Hg^{2+} and (B) HSO_4^-.

(a)   (b)

Through-bond   Through-space
communication   communication

Redox-active
group

2.62   **2.63**

(c)

2.64

Figure 2.58 Receptor-reporter electrochemical ion sensors functioning via the (A) Through-bond; (B) Through-space; and (C) Conformational change mechanisms.

visible to the naked eye is especially convenient. This is nicely demonstrated using the bis-indole receptor **2.61** (Figure 2.57B), which shows an obvious colour change from colourless to bright pink due to hydrogen bonding with HSO_4^- in acetonitrile/water 1:1 v/v.

Electrochemical ion sensing is highly advantageous from a practical point of view due to its high sensitivity, and facile integration of redox sensors into electronic devices for convenient sensing applications. Changes in redox properties of the receptor can be detected by voltammetric techniques such as cyclic voltammetry (CV) or square-wave voltammetry (SWV), whose $E_{1/2}$ values are perturbed during ion binding. Receptor-reporter electrochemical ion sensors can function by numerous pathways. In *through-bond* interactions (Figure 2.58A), ion binding is communicated electronically across conjugated pathways to the redox centre. In contrast, *through-space* interactions (Figure 2.58B) occur largely via differences in electrostatic interactions between the ion and the redox group upon oxidation/reduction. Finally, ion binding can also induce conformational changes to the receptor that perturbs its redox potential. For example, binding of K^+ by receptor **2.64** (Figure 2.58C) induces a change in conformation where the cation is bound in between both crown ether moieties in a 1:1 sandwich complex. This redirects the lone pairs on the sulfur atoms adjacent to the ferrocene redox centre and causes a detectable cathodic shift in the ferrocene/ ferrocenium redox couple.

Other than receptor-reporters, IDAs and chemodosimeters are also highly effective for ion sensing, albeit less widely applied. Eric Anslyn designed an IDA approach for sensing of polyanionic inositol-1,4,5-triphosphate (IP_3), which plays important roles in biological signalling pathways. Initially, the dye 5-carboxyfluorescein was bound to polyguanidinium tripodal receptor **2.65** (Figure 2.59A). As the C_{3v} symmetry of **2.65** is more complementary with the similar shape of IP_3 than the dye, IP_3 was able to bind more strongly to the receptor and displace the dye from the binding pocket. This resulted in a significant loss of fluorescent intensity.

Electrochemically active receptors bearing Lewis acidic sites such as receptor **2.66** (Figure 2.59B) can exhibit very large changes to their redox couples when an anion

(a)

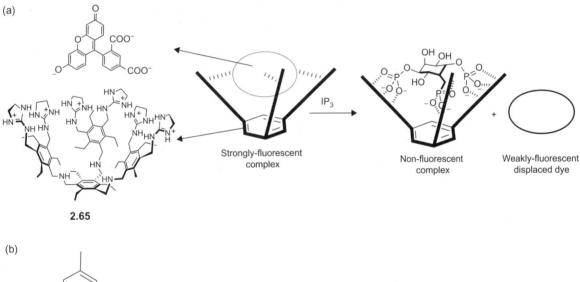

2.65

Strongly-fluorescent complex

Non-fluorescent complex

Weakly-fluorescent displaced dye

(b)

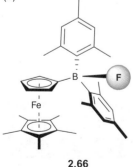

2.66

Figure 2.59 (A) An IDA fluorescent 'turn off' sensor for IP_3 (K. Niikura *et. al., J. Am. Chem. Soc.*, 1998, **120**, 8533); (B) redox and optical sensing of F^- by the chemodosimeter approach. (A.E.J. Broomsgrove *et. al., Chem. Eur. J.*, 2008, **14**, 7525)

coordinates directly onto the boronic acid. Both F⁻ and CN⁻ binding result in cathodic shifts of greater than 500 mV for the ferrocene/ferrocenium redox couples, as well as detectable visible colour changes of the complexes in a solvent comprised primarily of acetonitrile.

Biomedical applications

The ubiquity of ions in biological processes and our ability to design receptors for them opens up many possibilities of human intervention to treat diseases and regulate the activities of biomolecules. Other than imaging by optical ion sensors, cation/anion receptors can also regulate ion transport across cell membranes. Lipid bilayer membranes are inherently impermeable to ions due to their hydrophobic interior, but contain natural transport proteins which control their permeability towards these hydrophilic species. Sometimes, ion transport may be misregulated, which can lead to diseases. Cystic fibrosis for instance, has been attributed to misregulation of Cl⁻ transport. On the other hand, cation transport often has antibiotic applications as they can disrupt the natural transmembrane ionic gradients. Ion transport can be performed by synthetic receptors, such as macrobicycle **2.67** (Figure 2.60A), which function as mobile carriers that can co-transport KCl or NaCl across vesicle membranes. The contact ion-pair transport of **2.67** minimizes the polarity of the salt-bound complex, which enhances partitioning into the hydrophobic membrane interior. Alternatively, receptor **2.68** (Figure 2.60B) illustrates another strategy for transmembrane ion transport. Through π–π stacking self-assembly interactions, receptor **2.68** forms an ordered structure containing an ion-channel spanning the thickness of the membrane that allows Cl⁻ to pass through.

More recently, high-affinity host–guest recognition has enabled on-demand control of biomolecular activity. For example, the activity of bovine carbonic anhydrase enzyme is blocked by the benzenesulfonamide motif of an inhibitor compound (Scheme 2.3). As CB[7] forms a very stable 1:1 host–guest complex with the ammonium adamantyl unit (see Section 2.4 Receptors for binding organic cations) of the same inhibitor molecule, its addition causes the inhibitor to be released from the enzyme's

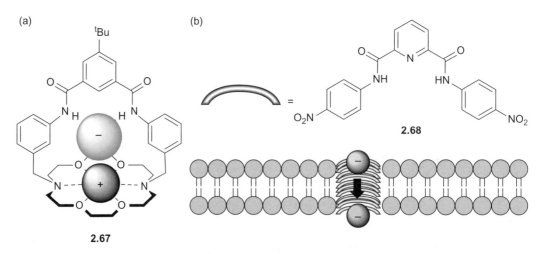

Figure 2.60 (A) A macrobicyclic ion-pair transmembrane transporter **2.67** for KCl and NaCl. (B) Self-assembly of **2.68** to form an ion-channel allowing Cl⁻ transport.

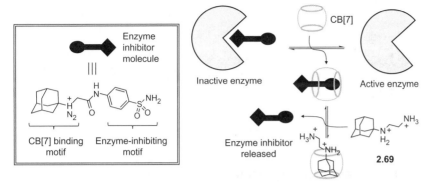

Scheme 2.3 Control of enzyme activity using CB[7] and adamantyl derivatives.

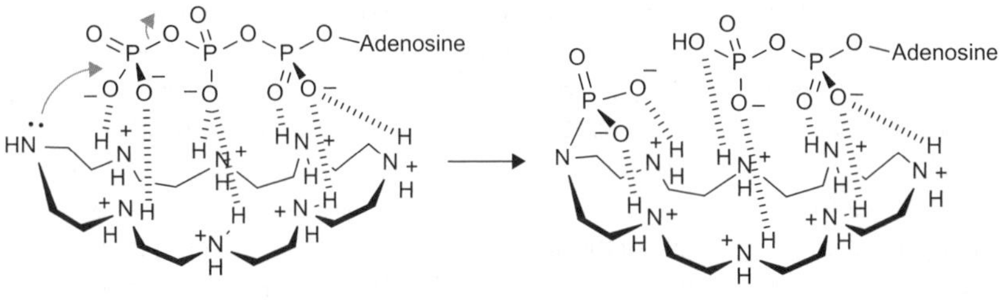

2.70

Scheme 2.4 Catalytic hydrolysis of ATP^{4-} by polyammonium macrocycle **2.70**.

active site and restores the enzymatic activity. However, the presence of diamine **2.69**, which forms a more stable association with CB[7] than the inhibitor molecule, can displace the inhibitor from CB[7]. The released inhibitor is now free to bind to the enzyme's active site again to negate its activity.

Catalysis

Supramolecular catalysis has witnessed great advances in recent years. The possibility of using ion-binding receptors for catalysis was realized in the early 1990s by Lehn. In a now classic example, macrocyclic polyamine **2.70** was demonstrated to effect an over 300-fold enhancement in the rate of ATP^{4-} hydrolysis at pH 3.5 (Scheme 2.4). Detailed mechanistic studies established the importance of ATP^{4-} binding to the multiply protonated receptor for hydrolysis. The host–guest complex possesses a similar structure to the transition state of the reaction, reducing its entropic penalty. Furthermore, charge neutralization of the polyanionic ATP^{4-} upon complex formation also facilitates attack by electron rich nucleophiles, as well as stabilizing the developing negative charges on the terminal phosphate group to enhance leaving group efficacy.

Other than affecting the rate of the reaction, ion-binding receptors can also alter the reaction outcome and resulting product distribution. For instance, to control the enantioselectivity of a Pictet–Spengler-type ring-closing reaction (Scheme 2.5), chiral thiourea receptor **2.71** is added as an organocatalyst which binds to the $(CH_3)_3SiO^-$ leaving group. This forms a chiral anionic complex which associates closely with the cationic iminium intermediate, creating an asymmetric environment that allows cyclization to occur with excellent stereoselectivity (up to 97% ee).

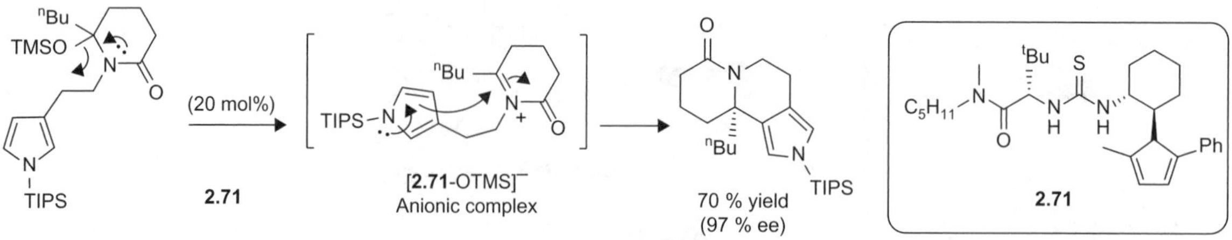

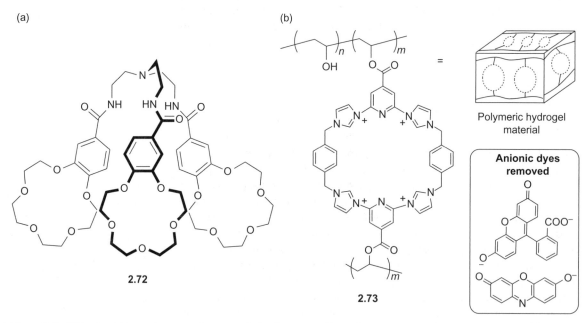

Scheme 2.5 Control of enantioselectivity with chiral anion-binding catalyst. (I.T. Raheem *et. al.*, *Org. Lett.*, 2008, **10**, 1577)

Extraction and recovery of ions

Ion extraction is useful for removing toxic salts from aqueous waste effluents, as well as recovering precious metal cations or anions from crude ores or waste streams. The pertechnetate anion ($^{99}TcO_4^-$) is radioactive and constitutes a major water pollutant from nuclear fuel processing plants. To remove the radioactive anion from waste effluents, tripodal receptor **2.72** (Figure 2.61A) was designed as an ion-pair receptor to extract TcO_4^- as its sodium salt. **2.72** displayed positive cooperativity for anion extraction in the presence of Na^+: one equivalent of the cation can boost anion binding affinity by 20-fold due to Na^+-mediated receptor reorganization as well as boosting the hydrogen-bonding acidity of the amide protons. To remove anionic pollutants for water remediation, the group of Jonathan Sessler has designed a crosslinked polymeric hydrogel containing cationic macrocycles which can interact with and remove a range of inorganic anions (F^-, NO_2^-, SH^-, HSO_4^-) as well as anionic dyes (Figure 2.61B) mainly by hydrogen bonding and electrostatic interactions. The pollutant-impregnated hydrogel polymer material can be removed from the aqueous solution and recycled

(a)

(b)

2.72

2.73

Polymeric hydrogel material

Anionic dyes removed

Figure 2.61 (A) A tripodal ion-pair receptor for removal of radioactive $Na^{99}TcO_4$ from aqueous nuclear waste discharges. (B) A cationic macrocycle-containing polymeric hydrogel for remediation of inorganic anions and dyes.

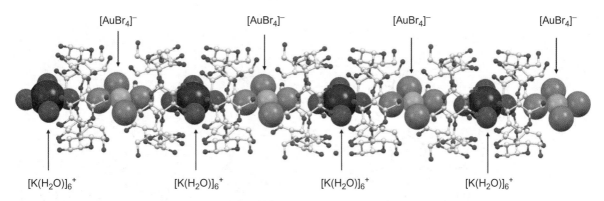

$[AuBr_4]^-$ $[AuBr_4]^-$ $[AuBr_4]^-$ $[AuBr_4]^-$

$[K(H_2O)]_6^+$ $[K(H_2O)]_6^+$ $[K(H_2O)]_6^+$ $[K(H_2O)]_6^+$

Figure 2.62 Crystal structure of the adduct between α-CD, $AuBr_4^-$, and $K(H_2O)_6^+$ used as a basis for a patented gold recovery process less harmful to the environment. (Crystal structure refcode: CSD-YIHJUW, first published: Z. Liu *et. al., Nat. Commun.*, 2013, **4**, 1855.)

for repeated use by simply soaking in excess dilute hydrochloric acid, where excess Cl^- displaces the bound pollutant anions and dyes.

In 2013, the group of Stoddart discovered serendipitously that α-cyclodextrin (α-CD) macrocycles were able to form high-yielding needle-like crystals within minutes with aqueous $KAuBr_4$. A detailed study revealed that the crystals comprised of a 2:1 stoichiometric outer-sphere adduct of α-CD, $AuBr_4^-$, and $K(H_2O)_6^+$ joined repeatedly in a one-dimensional superstructure (Figure 2.62). The ability of α-CD to extract $KAuBr_4$ is highly selective, and does not occur with complex anions of other transition metals such as Cu, Zn, Pd, Pt, and Ag. On the basis of this exquisite selectivity, a method to recover gold from crude ore has been developed, using the environmentally benign naturally occurring α-CD to replace toxic cyanide salts traditionally used in gold mining.

Supramolecular materials

Host–guest chemistry based on the recognition of ions can provide fertile ground for the imaginative design of materials with unusual, yet valuable properties. The stringent linearity of halogen bonding, for instance, can be exploited to form large rod-like supramolecular anions where an iodide anion is bound by two iodo-terminated linear fluorocarbons (Figure 2.63A). Interestingly, these 'super-anions'

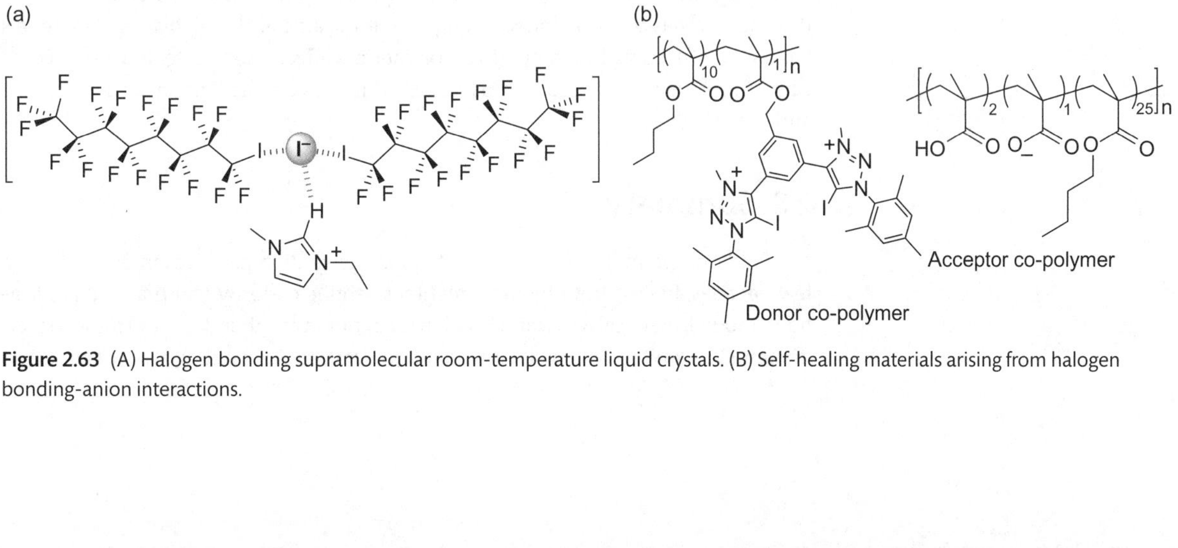

(a)

(b)

Acceptor co-polymer

Donor co-polymer

Figure 2.63 (A) Halogen bonding supramolecular room-temperature liquid crystals. (B) Self-healing materials arising from halogen bonding-anion interactions.

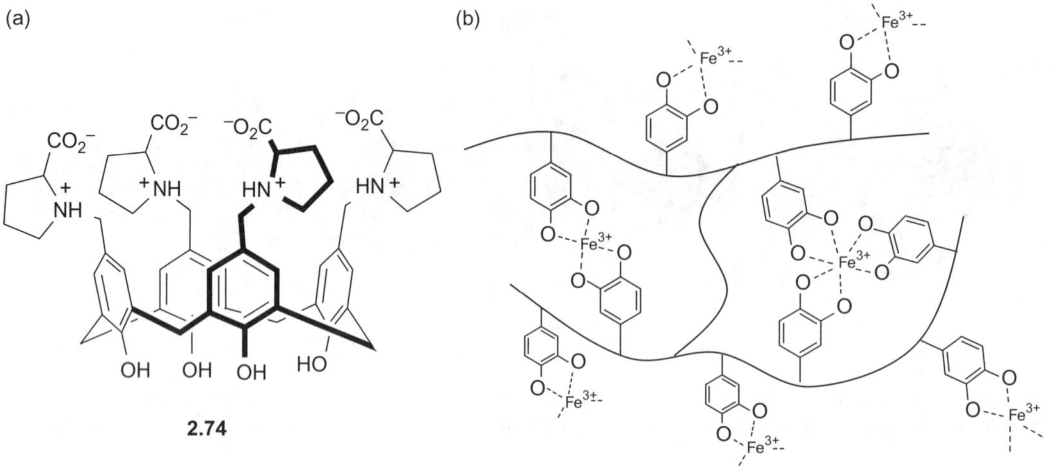

(a)

(b)

Figure 2.64 (A) Proline-functionalized calix[4]arene capable of forming hydrogels with hydrophobic anions. (B) Tough polymeric network hydrogels held together by strong and stable catechol-Fe^{3+} coordination.

were able to demonstrate the rare phenomenon of forming liquid crystals at room temperature. Separately, halogen bonding–anion interactions have proven instrumental in the design of self-healing supramolecular polymers. Blends of a halogen bonding cationic 'donor' co-polymer and an anionic 'acceptor' co-polymer (Figure 2.63B) formed a stiff and hard film. By virtue of the reversible halogen-bonding interactions between both polymeric components, physical damage of the film by scratching could self-heal with the application of heat due to the mutual re-assembly of both components.

Supramolecular receptor–ion interactions are very useful in engineering three-dimensional molecular network hydrogel materials, capable of trapping large quantities of water. Hydrogels are a very important class of soft materials with widespread recent applications in tissue engineering and regeneration, drug delivery, and molecular sensing. Proline-functionalized calixarene **2.74** (Figure 2.64A) forms hydrogels with hydrophobic anions such as I^-, ClO_4^-, and NO_3^- but not with strongly hydrophilic anions like SO_4^{2-}. In addition, the ability of catechol groups to form very stable complexes with Fe^{3+} (as we have seen for siderophores in Section 2.4 earlier in this chapter) has been exploited to engineer mechanically tough hydrogel materials (Figure 2.64B). Using biocompatible polymeric scaffolds, such catechol-Fe^{3+} hydrogels have shown great promise as tissue adhesives for facilitating wound closure and recovery.

2.8 Summary

The field of supramolecular cation and anion recognition has blossomed in the past few decades, fuelled both by fundamental scientific curiosity and practical applications. Indeed, practical applications of ion recognition are abundant, and future developments are limited only by the imagination. Despite the dazzling array of countless ion receptors developed, host–guest complementarity and receptor preorganizing are recurring leitmotifs throughout, governing both guest affinity and selectivity. The following flowchart provides a general guide on receptor design based on the properties of the target guest species:

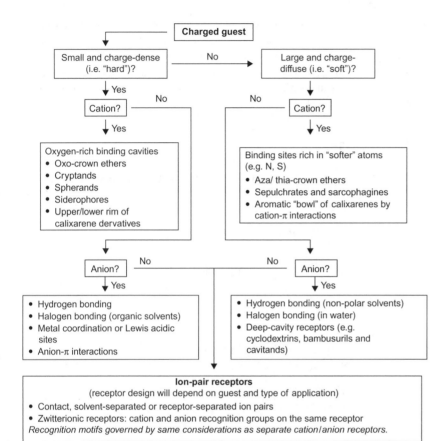

2.9 Further reading

Owing to the page limitations of the primer, we are only able to provide a brief glimpse of the immense developments in the ion recognition field achieved in the past decades. Fortunately, the literature is rich with more in-depth treatments of various aspects of cation/anion binding:

- For a general coverage of the entire field of anion receptor chemistry, Sessler, J. L., Gale, P., and Cho, W.-S., 2006, *Anion Receptor Chemistry*, Cambridge: RSC Publishing, is an excellent reference. In addition, *Chem.*, 2016, **1**, 351–422 provides an update on the recent developments of the field and *Chem. Rev.*, 2015, **115**, 8038–155 gives an in-depth discussion on applications of anion binding.

- A personal account of many key researchers in the development of supramolecular cation and anion receptors can be found in the book: Izatt, R. M. (Ed.), 2006, *Macrocyclic and Supramolecular Chemistry: How Izatt–Christensen Award Winners Shaped the Field*, Chichester: Wiley, particularly in Chapters 3, 5, 6, 8, 13, 14, 17, 18, 20, and 21.

- Anion recognition, sensing, and other applications using halogen bonds are reviewed in: *Chem. Commun.*, 2016, 52, 8645–58. Halogen bonding is a member of a family of supramolecular interactions known as sigma-hole interactions that is widely expected to play important roles in the field. Recent developments in the design and applications of anion binding by sigma-hole receptors are reviewed in: *Chem.*, 2018, **4**, 731–83.

2.10 Exercises

2.1 Consider the binding data for receptors **L₁**, **L₂**, and **L₃** in methanol: water 9:1 (*T* = 293 K).

Anion	L_1 K (M^{-1})	L_2 K (M^{-1})	L_3 K (M^{-1})
Cl$^-$	<10	<10	20
Br$^-$	28 800	631 000	85
I$^-$	955 000	37 100	80

(a) Why is Cl$^-$ most weakly bound in this solvent medium?

(b) For **L₁**, I$^-$ binds much more strongly than Br$^-$. However, Br$^-$ binds more strongly than I$^-$ to receptor **L₂**. Suggest a possible reason for this.

(c) If the atoms labelled X in the receptor were replaced with chlorine, predict the expected binding constant for I$^-$ in the same solvent.

2.2 Binding constants K (M^{-1}) for **L₄** and anions in the absence and presence of NaClO₄ in acetonitrile (CH₃CN) solution are shown below:

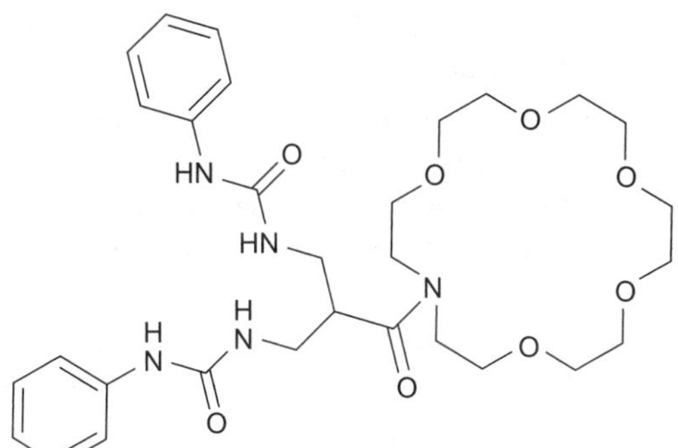

L_4

Anion	L_4 K (M^{-1})	L_4 + NaClO$_4$ K (M^{-1})
Cl$^-$	18 400	53 000
Br$^-$	770	3900
NO$_2^-$	2100	5800
PhCOO$^-$	242 000	251 000

(a) Suggest why a smaller cooperativity factor for Cl$^-$ was seen compared to Br$^-$ in the presence of NaClO$_4$.

(b) Would you expect NO$_3^-$ to bind more strongly or weakly than NO$_2^-$ in the absence of NaClO$_4$?

(c) Explain why the receptor is selective for benzoate compared with the other anions. Why is the cooperativity factor for benzoate in the presence of NaClO$_4$ so close to 1?

2.3 Chalcogen bonding is a σ-hole interaction considered to be a sister supramolecular interaction to halogen bonding due to their many similarities. Unlike halogen bonding, whose donor atoms are Group 17 elements (Br/ I), the donor atoms for chalcogen bonding are those from Group 16 (S/ Se/ Te). Only since 2015 has the potential for chalcogen bonding in anion recognition been explored in earnest with the design of abiotic chalcogen-bonding receptors capable of binding anions in solution. The following shows a possible family of chalcogen bonding hosts for anions:

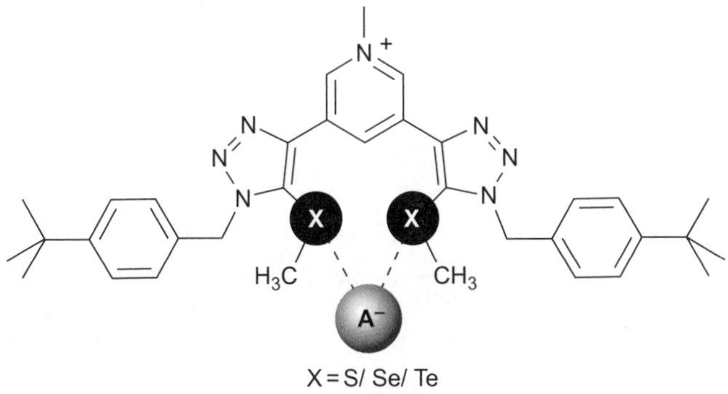

X = S/ Se/ Te

(a) Predict how the Cl⁻ binding affinity of the S, Se, and Te-containing family of chalcogen bonding hosts would differ in acetone.

(b) The presence of the cationic pyridinium group can attract anions electrostatically and compete with the chalcogen-bonding binding site for coordination of anions. Explain how you would determine the predominant location of anion binding on the host molecule spectroscopically.

(c) In acetonitrile, the binding affinity of the Te-containing host for Cl⁻ was found to decrease as temperature increases: $K_a = 652$ M^{-1} (298 K); 484 M^{-1} (313 K); 410 M^{-1} (323 K) and 356 M^{-1} (338 K). Calculate ΔG, ΔH, and $T\Delta S$ for Cl⁻ binding at 298 K.

For the answers to these exercises, visit the online resources which accompany this primer.

Binding of neutral guests

3

3.1 Introduction

A large number of molecules that we encounter in our daily lives are uncharged. These include, but are not limited to, drugs which we take to treat our maladies, vitamins essential for maintaining health, scents and volatile organic compounds, as well as fragrances and bioactive ingredients which are common additives to body washes, shampoos, and lotions. As we shall see in the course of this chapter, there are very good reasons for binding these neutral molecules, and important supramolecular host–guest recognition strategies underpin many formulations found on supermarket shelves. Unlike binding of ions, considered in Chapter 2, the binding of neutral molecules necessitates different strategies. Primarily, due to their uncharged nature, the interactions between neutral guests and host molecules are comparatively a lot weaker than the dominant electrostatic forces of ions. Instead, receptors designed to bind neutral molecules have to rely on hydrogen bonding, π–π interactions, Lewis acid interactions, and the hydrophobic effect to a much greater extent. As we shall see in Section 3.2 of this chapter, selectivity by dedicated host molecules is now dictated to an even greater extent by the fundamental principles of complementarity and preorganization discussed in Chapter 1. In addition, binding by 'physical incarceration/ imprisonment' of neutral molecules within the cavities of larger host molecules (Section 3.3) now plays a much bigger and prominent role than that for ions. Finally, we shall briefly consider in Section 3.4 how macroscopic crystalline solid host molecules, such as **metal–organic frameworks** (MOFs) can be exploited and engineered for binding of neutral guest molecules both in solution and in the gaseous phase.

3.2 Host–guest recognition: rational design of hosts for neutral guest binding

Directional guest binding by hydrogen bonding receptors

Hydrogen bonding is arguably the most abundant supramolecular interaction in biology. For optimal binding, the interacting hydrogen bond donor and acceptor groups must be perfectly aligned, such that the interactions between them are as linear as possible to maximize their strength. *Arrays* of multiple hydrogen bonds are also often present to further augment the binding strength. One of the strongest non-covalent

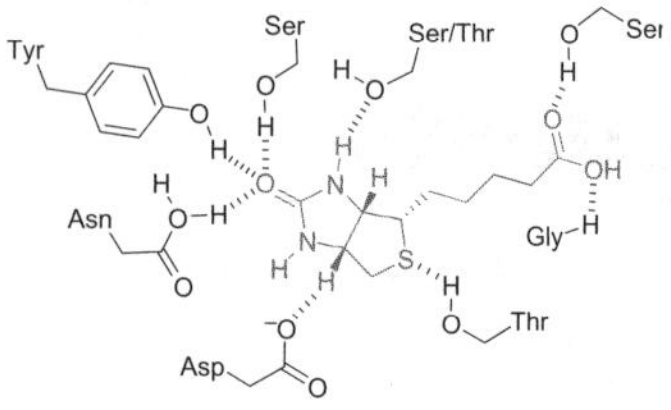

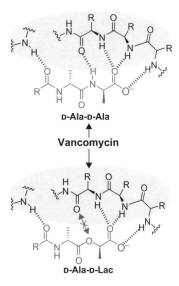

Figure 3.1 Biotin binding within the hydrophobic binding pocket of Streptavidin by a hydrogen bonding array.

D-Ala-D-Ala

↑

Vancomycin

D-Ala-D-Lac

Figure 3.2 Hydrogen bonding arrays accounting for the difference in binding affinities between D-Ala-D-Ala and D-Ala-D-lactate with vancomycin.

host–guest binding pairs found in nature occurs between the bacterial protein Streptavidin and biotin (otherwise known as Vitamin B7), with an impressive binding constant of approximately 10^{14} M^{-1}. As shown in Figure 3.1, biotin binds within Streptavidin's dehydrated and hydrophobic binding pocket, allowing eight strong and geometrically optimal hydrogen bonds to form with surrounding amino acid residues.

Another well-known example highlights how important host–guest hydrogen bonding complementarity can be for interactions between biological molecules. The antibiotic vancomycin works by forming a hydrogen-bonded complex with a peptide terminated by a D-Ala-D-Ala sequence (Figure 3.2), which inhibits its natural function to synthesize the bacterial cell wall. When one amide group is replaced with an ester in D-Ala-D-lactate, the loss of just one hydrogen bond reduces the peptide's affinity to vancomycin by almost 3 orders of magnitude, resulting in antibiotic resistance. Even more impressively, disruption of the host–guest binding geometry by replacing the terminal D-Ala residue with L-Ala completely eliminates any host–guest binding.

The design of synthetic receptors also employs many of the key concepts adopted by biological molecules. Consider the class of molecules called barbiturates, which are a useful class of anti-convulsant and sedative drugs. Figure 3.3A depicts how one may design a potential host molecule for this class of compounds from the bottom-up by considering the structure of the barbiturate guest. Firstly, the alternating hydrogen bond acceptor (C=O) and donor (N–H) groups on barbiturate **3.1** requires the *complementary* alternating donor (D) and acceptor (A) groups on a potential receptor respectively. Such a binding motif may be found in pyridine 2,6-substituted with amide groups, where the pyridine nitrogen atom provides the hydrogen bonding acceptor unit and the amide the donor. Secondly, the C_2-symmetry of **3.1** requires a similar host symmetry with two such pyridine-2,6-amide groups separated by an appropriate spacer motif. Thirdly, the binding motif on the host should be rigid to prevent self-dimerization. Finally, the host should be preorganized to maximize binding affinity, which can be achieved through the incorporation of the host binding motif into an appropriately sized macrocycle. Indeed, Andy Hamilton designed a macrocyclic receptor for barbiturates fulfilling all the above criteria in the 1980s. In his receptor **3.2**, the barbiturate is held by six convergent and complementary hydrogen bonds to the rigid binding motif. In non-polar dichloromethane, **3.2** binds barbiturate **3.1** with an impressive binding affinity of 250 000 M^{-1}. When two carbonyl groups are removed from **3.1**, the loss of two host–guest hydrogen bonds reduces the resulting urea's binding affinity to just 400 M^{-1} in the same solvent.

In 2019, a receptor capable of binding D-glucose impressively strongly in water was reported. Designed from first principles by considering the molecular properties of the guest (six-membered β-pyranose form of D-glucose **3.3**), this receptor is an example of 'bottom-up' receptor design par excellence. As shown in Figure 3.4A, the β-pyranose form of D-glucose possesses an all-equatorial arrangement of polar hydroxyl groups, with two hydrophobic vicinities on the top and bottom face of the molecule comprising of the axial C–H units. To complement the positions of these hydrophobic areas, trimethylmesitylene units are positioned at the 'roof' and 'floor' of receptor **3.4**, (Figure 3.4B), allowing hydrophobic C–H···π interactions with D-glucose. The six-membered pyranose ring of glucose also necessitated a receptor with a threefold symmetry. Furthermore, each arm comprising a polar bis-urea unit is twisted out of the plane from each other, allowing each urea unit to form hydrogen bonds with the equatorial vicinal OH groups of the guest. By virtue

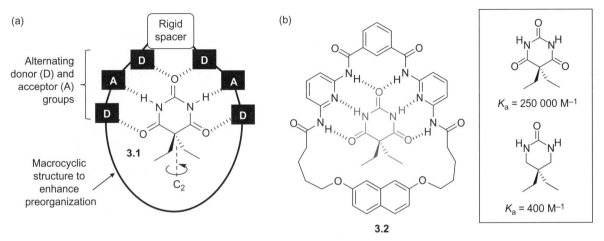

Figure 3.3 (A) Bottom-up design of a receptor for barbiturate **3.1**; (B) Hamilton's barbiturate-binding macrocycle **3.2** with six complementary hydrogen bonds. (S. K. Chang & A. D. Hamilton, *J. Am. Chem. Soc.*, 1988, **110**, 4, 1318).

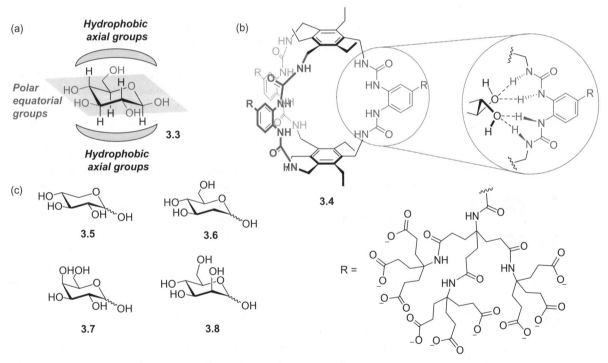

Figure 3.4 (A) Molecular properties of glucose in its six-membered β-pyranose form; (B) structure of Anthony Davis' glucose receptor; (C) structures and binding affinities of related hexoses. D-xylose (**3.5**), 2-deoxy-D-glucose (**3.6**), D-galactose (**3.7**) and D-mannose (**3.8**). (R.A. Tromans *et. al.*, *Nat. Chem.*, 2019, **11**, 52)

of its excellent complementary design, the receptor exhibits a very high degree of selectivity for glucose binding over other structurally related pyranose sugars. For instance, the structurally related D-xylose (**3.5**), 2-deoxy-D-glucose (**3.6**), D-galactose (**3.7**) and D-mannose (**3.8**) are all bound considerably more weakly (Figure 3.4C). It

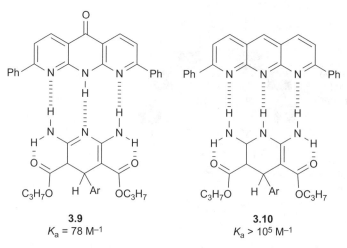

3.9
$K_a = 78\ M^{-1}$

3.10
$K_a > 10^5\ M^{-1}$

Figure 3.5 Stability constants of two host–guest complexes held by DAD–ADA and DDD–AAA hydrogen bonding arrays. (T.J. Murray & S.C. Zimmerman, *J. Am. Chem. Soc.*, 1992, **114**, 4010)

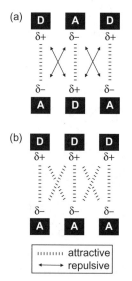

Figure 3.6 How secondary diagonal dipolar interactions can weaken and strengthen the stability of a (A) DAD–ADA and (B) DDD–AAA hydrogen bonding arrays, respectively.

is especially remarkable that simply exchanging the equatorial C4- and C2-OHs of glucose to the axial positions on D-galactose and D-mannose results in a more than two orders-of-magnitude reduction in K_a.

Although host–guest binding using an *array* of hydrogen bonds can confer strong association, there exist secondary interactions which can greatly influence the stability of the complex. For instance, consider the host–guest complexes **3.9** and **3.10** (Figure 3.5), each held together by three linear hydrogen bonds but differing greatly in stability. To explain why complex **3.9** binds so much more weakly, we note that the host molecule (top) comprises of a acceptor-donor-acceptor (ADA) fragment, and the guest (bottom) has the complementary donor-acceptor-donor (DAD) fragment. As hydrogen bonds have significant *electrostatic* character, resulting from an electropositive hydrogen atom and an electronegative acceptor atom, *diagonal* electrostatic dipolar interactions between adjacent hydrogen bonds within the closely spaced array become important. In the DAD–ADA array of **3.9**, while the individual linear hydrogen bonds are attractive, the diagonal electrostatic dipole attractions are repulsive as they arise from centres with like dipole charges (Figure 3.6A). These diagonal interactions have in fact been shown by calculations to be approximately one-third the strength of the primary linear hydrogen bonds! Contrastingly, the DDD–AAA array of complex **3.10** are all attractive (Figure 3.6B), accounting for its much greater stability.

Hydrogen bonding arrays are important constituents in practical supramolecular polymer networks. Ureido-pyrimidinone (UPy) is a popular DDAA–AADD self-associating motif, capable of dimerizing with itself by forming a quadruple hydrogen bonding array (Figure 3.7). Contributing to its self-dimerization affinity is the stable intramolecular hydrogen bond which 'locks' the conformation of the urea motif with the adjacent pyrimidinone group in a stable six-membered ring, preventing its free rotation. By appending the UPy group onto biocompatible polymer fragments such as poly(ethylene glycol) (PEG), their self-association in water can form supramolecular polymer networks and hydrogels, which have been studied extensively for delivery and controlled release of therapeutic drugs under numerous clinical settings.

Exploiting π–π stacking interactions

While hydrogen bonding confers selectivity and organization to guest binding by molecular hosts, donor-acceptor π–π stacking interactions can augment the stability of host-guest complexes significantly. Receptor **3.11** is an interesting host molecule which binds a mimic of the neurotransmitter serotonin **3.12** (Figure 3.8). Other than a combination of hydrogen bonds, including two unusual ones between N–H and C–H donors on the edge of the guest with the electron rich face of the anthracene moiety, host–guest stabilization is augmented by π–π stacking interactions between **3.12** and the receptor's top and bottom naphthalene units.

An emerging class of nanocarbon materials are fullerenes, which are allotropes of carbon (e.g. C_{60} and C_{70}). Donor–acceptor interactions between porphyrins and fullerenes have been intensively studied in recent years due to the possibility of modulating their electronic and optical properties for light harvesting, artificial photosynthesis, and medical imaging applications. While π–π interactions are traditionally thought to be limited to flat surfaces between host and guest molecules, the flat π surface of porphyrins and the curved π surface of fullerenes can still bring about impressive binding strengths. The receptor **3.13** (Figure 3.9A) consists of two zinc-porphyrin motifs appended onto a spacer group preorganized by intramolecular hydrogen bonds, such that the receptor adopts a rigid U-shaped conformation. As a result, C_{60} is bound between both porphyrin units of the molecular tweezer by π–π interactions with binding constants up to 10^5 M^{-1} in toluene. These interactions also feature in 3D molecular cages flanked by porphyrins with cavities designed to bind fullerenes inside them. The cage **3.14** (Figure 3.9B) can bind C_{60} and C_{70} very strongly and can be used to isolate and purify fullerenes from the complex multicomponent mixtures.

Boronic esters

In Chapter 2, we saw how Lewis acidic boron atoms in receptors can be used to bind and sense nucleophilic anions such as F$^-$ and CN$^-$. Boron, in the form of boronic acids, is also supremely useful for binding compounds with the 1,2-diol motif. As shown in

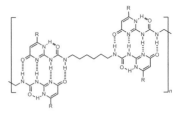

Figure 3.7 The DDAA-AADD hydrogen bonding array between a dimer of the popular UPy motif used in supramolecular polymers.

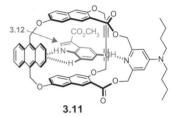

Figure 3.8 Binding of serotonin mimic **3.12** using receptor **3.11** via a combination of hydrogen bonding and π–π stacking interactions.

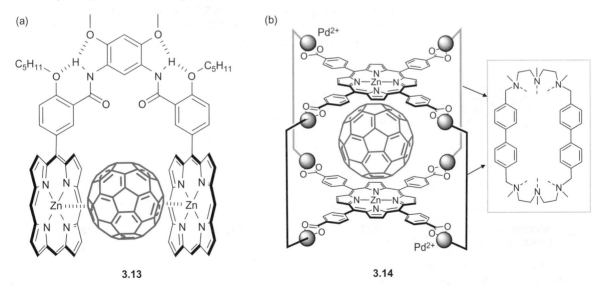

Figure 3.9 (A) A U-shaped molecular tweezer for C_{60} binding; (B) a molecular cage for C_{60} binding and separation from a fullerene mixture.

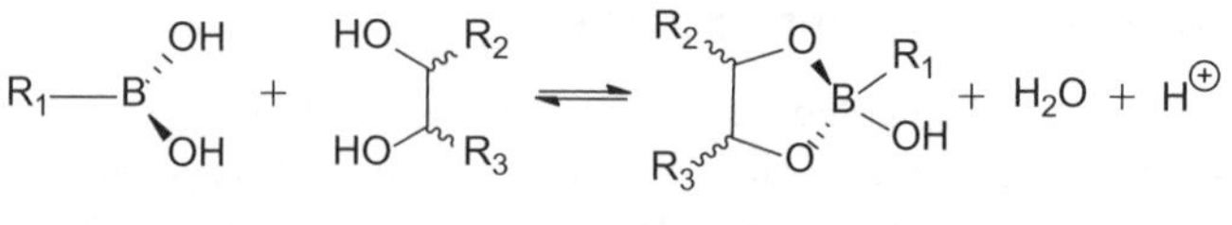

Scheme 3.1 Rapid and reversible formation of cyclic boronic esters from boronic acids using 1,2-diols.

Scheme 3.1, the trigonal planar boronic acid **3.15** can rapidly and reversibly form a stable five-membered cyclic boronic ester **3.16** with 1,2-diols even in competitive solvent media such as water, liberating H$^+$ and a water molecule. This changes the hybridization of the boron from sp^2 to sp^3, and its geometry from trigonal planar to tetrahedral. While it should be noted that the formation of boronic esters is a covalent reaction and not a supramolecular interaction, this reaction has found extensive use in the design of receptors for the recognition of sugars.

Diol motifs are commonly found in saccharides, and hence boronic acid-containing receptors have traditionally been employed for detecting sugars such as glucose. In 2001, a highly selective diboronic acid glucose receptor was designed with the aid of computation. The target mode of coordination to the six-membered form of glucose (termed glucopyranoside) was initially designed (**3.17** in Figure 3.10) with both boronic acid groups in the correct relative positions to interact with the appropriate α-1,2 and 4,6-hydroxy diol linkages on the guest. The lowest energy conformation of synthon **3.17** was obtained by computational geometry optimization. To identify the spacer motif represented by the large oval in Figure 3.10, a programme was then used to search for polycyclic organic structures with substituent bonds closely matching the optimized geometry of synthon **3.17**. Of the approximately 300 possible structures generated, compound **3.18** was chosen as the lead structure due to its apparent structural simplicity and ease of synthesis. This compound was then modified to obtain the final receptor structure **3.19**, replacing the terminal five-membered fulvene ring with the isosteric furan motif which was more chemically stable and easily synthesized. The value of this computer-aided design was proven when **3.19** bound glucose with K$_a$ > 10^4 M^{-1} in methanol-aqueous buffer solvent (pH 7.5), a more than 400-fold greater affinity than galactose, mannose, and fructose.

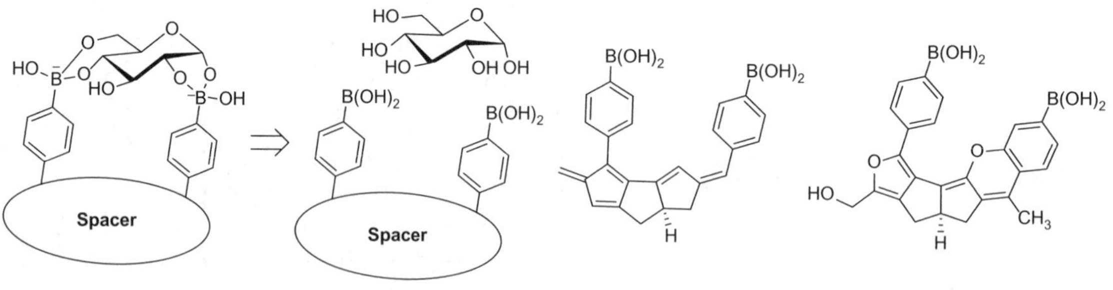

Figure 3.10 Stages in the computer-aided design of glucose receptor **3.19**. For more information, please see W. Yang et. al., Angew. Chem. Int. Ed., 2001, 40, 1714–1718.

Emerging applications

Supramolecular host–guest complexation within size-complementary cages can stabilize reactive molecules. A particularly impressive example involved binding white phosphorus (P_4) within a tetrahedral cage **3.20** (Figure 3.11). White phosphorus, comprising of a tetrahedral arrangement of four phosphorus atoms, spontaneously combusts when exposed to oxygen in the air. When encapsulated within the cage, P_4 is stabilized via hydrophobic van der Waals interactions with the phenylene units lining the cage's interior, and is completely stable to contact with the atmosphere for over 4 months. The origin of this remarkable stability arises not from the inability of O_2 to penetrate the cage, but from the fact that oxidation of P_4 forms intermediate products too large to fit within the rigid cage structure.

A very powerful concept for molecular sensing is called '***differential sensing***'. Instead of using a highly specific host molecule for a guest, an *array* of receptors is employed to generate a unique fingerprint sensing pattern or identification signature for each guest molecule, which is a culmination of each individual receptor's affinity for the guest. Advantageously, this approach eliminates the need to design and synthesize *specific* receptors for guests, as the same array can be used for a multitude of analytes. For example, an array of histidine-rich peptides, metal cations and colorimetric dyes was used to identify a number of flavonoids (Figure 3.12). 'Differential sensing' was applied to analyse wine, which comprises of a complex mixture of organic molecules, including neutral compounds such as flavonoids and organic acids (e.g. citric, malic, and tartaric acid). Targeting the flavonoids present, the same sensing array generated unique sensing fingerprints for each wine variety (e.g. Shiraz, Merlot, Pinot noir, Zinfandel, Beaujolais, and Cabernet sauvignon). This approach, which mimics how the sophisticated mammalian senses of taste and smell work, enabled the different wine varieties to be identified.

3.3 Guest binding within hydrophobic cavities

Like hydrophobic cations and anions (Chapter 2), neutral organic molecules can be bound within the hydrophobic cavities of macrocyclic host molecules. This is especially pronounced in water, augmented by the hydrophobic effect. A variety of such macrocycles with 'greasy' internal cavity walls are known, and these host–guest complexes are extremely useful for a large number of applications, as we shall see.

Cyclodextrins

Cyclodextrins (CDs) are naturally occurring water-soluble macrocycles containing D-glucopyranose units linked by α-1,4-glycosidic bonds. Produced by bacterial action on starch, the most abundant and widely studied members of this family are α, β, and γ-CD, comprising of 6, 7, and 8 glucose units respectively. As a result, these macrocycles possess different cavity sizes (Figure 3.13), which are useful in forming **inclusion complexes** with hydrophobic guest molecules of the appropriate dimensions. The reversible formation of these complexes is primarily driven by the hydrophobic effect, together with van der Waals interactions. CDs are regarded by the U.S. Food and Drug Administration (FDA) as safe, non-toxic substances for human consumption, and hence are very useful ingredients in many pharmaceutical formulations.

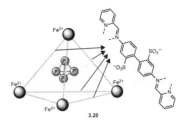

3.20

Figure 3.11 Jonathan Nitschke's tetrahedral cage for stabilizing reactive P_4 molecule from spontaneous combustion in air. The structure of each edge of the cage is shown separately for clarity.

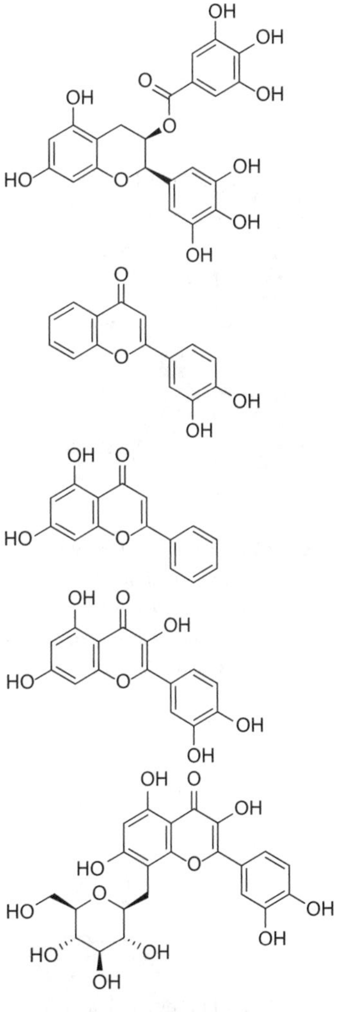

Figure 3.12 Structures of some flavonoid molecules found in wine.

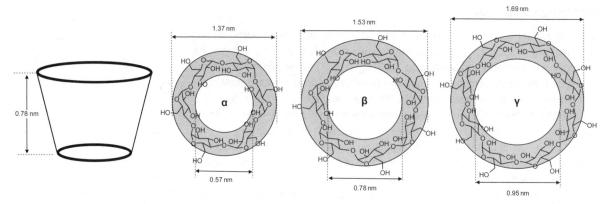

Figure 3.13 The dimensions and shapes of α, β, and γ-CD.

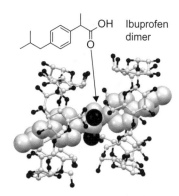

Ibuprofen dimer

Figure 3.14 Crystal structure of the inclusion complex of ibuprofen and β-CD. Crystal structure refcode: CSD-TUXKUS, first published: S. S. Braga et. al., *New J. Chem.*, 2003, **27**, 597.

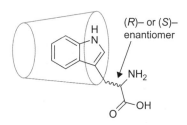

(R)– or (S)– enantiomer

Figure 3.15 Chiral discrimination between R- and S-tryptophan by α-CD.

In the pharmaceutical industry, CDs play a variety of hugely important roles. Firstly, many drug molecules are hydrophobic, and have poor solubility in water. As these drug molecules can only be absorbed by the body when dissolved, their inherent poor solubility reduces bioavailability. By forming inclusion complexes with CDs, the solubility of these drug molecules in water is greatly increased, sometimes by several orders of magnitude, which also increases their efficiency of uptake by the body. Ibuprofen, the anti-inflammatory pain reliever, is an example of a common drug whose water solubility is increased significantly when complexed with β-CD (Figure 3.14). Secondly, encapsulation by CDs can increase the stability of molecules that are prone to oxidation, rearrangement reactions, or decomposition. Acting as a physical barrier, CD encapsulation restricts the access of reactive molecules with the complexed guest. In addition, encapsulation disfavours rearrangement reactions as the different-shaped products are not of complementary geometry. Thirdly, CD encapsulation can prevent the binding of active pharmaceutical molecules to nasal or taste receptors, removing unpleasant sensations of bitterness or odours. Finally, many pharmaceutical ingredients are liquids at room temperature, and CD encapsulation can convert them into a powder form which simplifies handling, processing, and formulation procedures.

The excellent capabilities of cyclodextrins to form inclusion complexes with hydrophobic molecules have made them useful in formulations for air fresheners (e.g. Febreze®, which contains β-CD). When sprayed, the droplets containing β-CD readily form inclusion complexes with neutral unpleasant odour molecules in the air. Hence the β-CD molecules act as a molecular barrier to prevent these bad odour molecules from binding to olfactory receptors in our noses, preventing us from detecting them. Eventually, these inclusion complexes settle onto surfaces by gravity, to be easily removed.

Another property of cyclodextrins which is especially useful for molecular recognition is their inherent chirality. As we have seen, the glucopyranose units constituting CDs each possess five chiral centres. Remarkably, despite the many chiral centres (multiples of five) present, each CD molecule is homochiral, allowing the possibility of **chiral discrimination**, where one enantiomer is bound more strongly than the other. For instance, R-tryptophan binds more strongly in the cavity of α-CD in D_2O than S-tryptophan, where its indole ring can penetrate deeper into the host's hydrophobic cavity (Figure 3.15). Molecular modelling has shown that included R-tryptophan

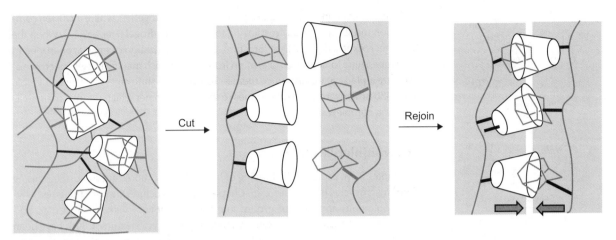

Figure 3.16 Self-healing hydrogels formed from inclusion complexation of adamantane into β-CDs.

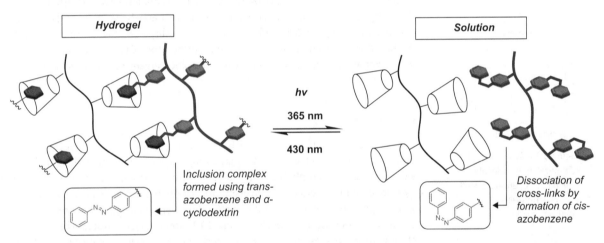

Figure 3.17 Stimuli-responsive α-CD-azobenzene hydrogels whose dimensions can be changed dynamically by irradiation using light of different wavelengths.

can form twice as many hydrogen bonds with the α-CD host compared to the *S*-enantiomer. Hence a further application for these host molecules is in chiral chromatography, where CDs bonded to the silica gel stationary phase of a chromatographic column can be used to separate enantiomers of amino acids and derivatives.

The imaginative exploitation of the reversible nature of CD host–guest complexation has enabled self-healing and stimuli-responsive **hydrogel** materials to be created, which has since blossomed into a field of its own. As shown in Figure 3.16, when acrylamide-functionalized inclusion complexes of β-CD and adamantane are polymerized *in situ*, a hydrogel forms. Physically cutting the hydrogel material forces the host and guest to decomplex at the point of fracture. By simply bringing the cut ends physically into contact with each other, the free β-CD and adamantane units can reform their inclusion complexes, restoring gel strength by 100 % to achieve perfect self-healing. To engineer hydrogels responsive to light, a polymer containing α-CD

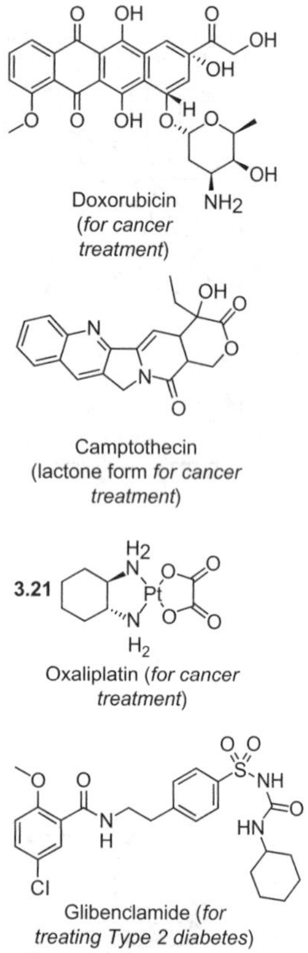

Doxorubicin
(for cancer treatment)

Camptothecin
(lactone form for cancer treatment)

3.21

Oxaliplatin *(for cancer treatment)*

Glibenclamide *(for treating Type 2 diabetes)*

Figure 3.18 Some examples of neutral drug molecules which have been successfully encapsulated within cucurbit[n]urils.

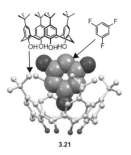

3.21

Figure 3.19 Crystal structure of a 1:1 complex of calix[4]arene **3.21** with 1,3,5-trifluorobenzene (H atoms not shown for clarity). Crystal structure refcode: CSD-EZUMER, first published: G. D. Enright *et. al., Chem. Commun.,* 2004, 1360.

and another containing trans-azobenzene were mixed together, where inclusion complexation results in physical crosslinking to form the hydrogel (Figure 3.17). When the hydrogel is exposed to UV light (365 nm), photoisomerization of trans-azobenzene to cis-azobenzene forces decomplexation from α-CD, which manifests itself macroscopically by a physical expansion of the gel. To cause gel shrinkage, exposure to visible light (430 nm) reverts the azobenzenes to their trans isomer which reforms their α-CD inclusion complexes.

Cucurbit[n]urils (CB[n])

In Chapter 2, we explored the binding of hydrophobic cationic guest molecules using CB[n] macrocycles (see Figure 2.25, Chapter 2, for structures), where the guest's hydrophobic units and cationic groups interact preferentially with the host macrocycle's cavity and portal oxygen atoms respectively. Like cyclodextrins, CB[n]s have been investigated for drug delivery by forming inclusion complexes with neutral hydrophobic drug molecules (Figure 3.18), which can enhance their aqueous solubilities, bioavailabilities, and stabilities towards the harsh pH conditions within the human gut. Toxicology studies on mice and on human cell lines have shown that CB[7] and CB[8] exhibit no toxicity up to 600 mg/kg, and a small quantity can actually be absorbed into the bloodstream from the alimentary canal. Furthermore, CB[7] does not accumulate in the liver, and can be quickly cleared from the bloodstream into urine without chemical modification.

Cyclophanes

Cyclophanes are a class of macrocycles containing at least one aromatic molecule bridged by aliphatic chains. The aromatic rings confer macrocycle rigidity and provide a restricted hydrophobic cavity suitable for binding guests by π–π, CH···π and hydrophobic interactions. The diversity of cyclophanes is extensive. Calix[n]arenes, whose applications in cation and anion binding were surveyed in Chapter 2, are an example of this class of molecules. Figure 3.19 shows a crystal structure of a 1:1 stoichiometric host–guest complex of a *p*-tert-butyl-calix[4]arene molecule **3.21** and 1,3,5-trifluorobenzene.

First reported in 2008 by Tomoki Ogoshi at Kyoto University, Japan, pillar[n]arenes are another class of cyclophanes closely related to calix[n]arenes, being comprised of n hydroquinone units joined in a cyclic structure by methylene bridges (Figure 3.20). Unlike calix[n]arenes, which adopt an open cup-like cavity, pillar[n]arenes adopt a cylindrical form. The interior of the pillar[n]arene cavities are π-electron rich, making them suitable for binding neutral guests with electron-withdrawing groups such as 1,2-dichloroethane or chlorocyclohexane. Like cyclodextrins, pillar[n]arenes can be useful for stimuli-responsive material engineering. For instance, a hydrogel containing ferrocene units attached onto a crosslinked polymer displayed dramatic physical swelling when anionic water-soluble pillar[6]arenes were added to the solution (Figure 3.21). Inclusion complex formation between the pillar[6]arenes and ferrocene generated strong electrostatic repulsion between the polymer chains, physically forcing them apart. When the pH of the aqueous solution was lowered, some of the anionic groups of the pillar[6]arenes were protonated, decreasing the extent of electrostatic repulsion to cause hydrogel shrinkage. When cationic drug molecules are loaded onto the hydrogel material, gel shrinkage upon acidification also concurrently triggered drug release.

    Unlike the π-electron rich cavities of pillar[n]arenes, a very important class of cyclophanes possessing π-electron deficient cavities were first reported by Nobel laureate J. Fraser Stoddart. Since the initial 'blue box' tetracationic cyclophane was reported (**3.22** in Figure 3.22A), named to represent its electron-deficient nature, a whole family of such macrocycles, possessing different sizes, shapes, and aromatic functionalities have been designed. The structure of cyclophane **3.22**, bearing two rigid 4,4-bipyridinium (paraquat) motifs, is most suited for binding electron rich planar aromatic guests such as tetrathiafulvalenes (Figure 3.22B). In acetonitrile for instance, **3.22** binds tetrathiafulvalene **3.23** with a K_a as high as 10 000 M^{-1} at 27 °C. A larger member of this family, **3.24** (ExBox2), forms an inclusion complex with C$_{60}$ (Figure 3.23). Interestingly, these inclusion complexes self-assemble to form one-dimensional needle-like crystals which exhibited electrical conductivity. The ability of these electron-deficient cyclophanes to form complexes with electron rich aromatic species using self-assembly (see Chapter 4) was exploited by Fraser Stoddart in pioneering work to construct mechanically interlocked structures such as rotaxanes and catenanes, which is discussed in Chapter 5. This class of cationic electron-deficient cyclophanes has also found numerous useful applications which include catalysis, molecular electronics, artificial photosystems, and molecular machines.

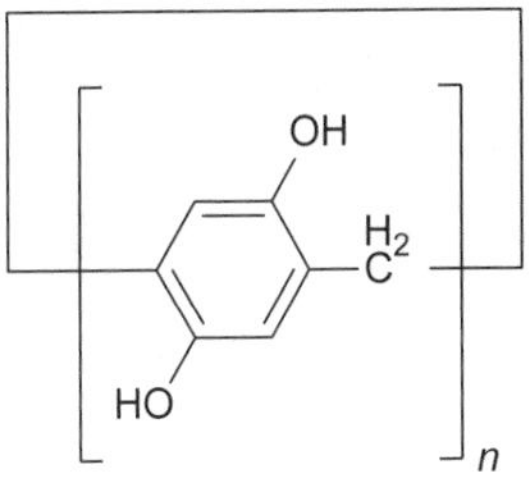

Figure 3.20 General structure of pillar[n]arenes.

Molecular containers with one opening: deep cavity cavitands

Unlike the cyclodextrins, cyclophanes, and curcubiturils we have seen so far that are toroidal and open at both ends, cavitands are host molecules which are shaped like a deep bowl—closed at one end but open at the other. Structurally, they comprise of a

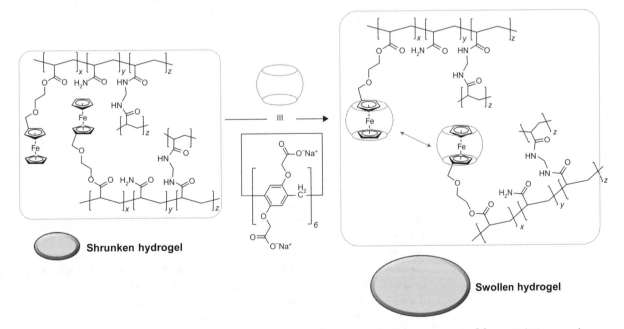

Shrunken hydrogel

Swollen hydrogel

Figure 3.21 Hydrogel expansion and pH-responsive drug release by a hydrogel containing ferrocene pillar[6]arene inclusion complexes.

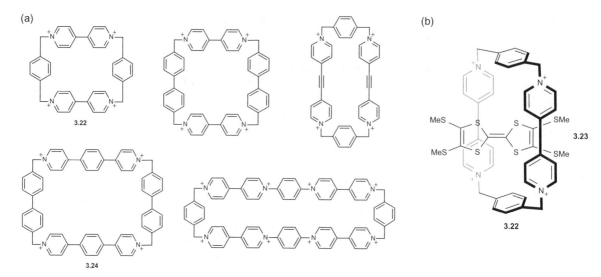

Figure 3.22 (A) The original 'blue box' cyclobis(paraquat-*p*-phenylene) cyclophane **3.22** and some examples of its extended family. (B) Host–guest complexation between **3.22** and a tetrathiafulvalene guest **3.23**.

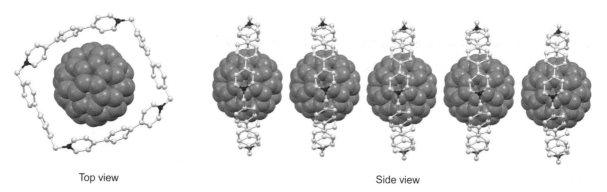

Top view   Side view

Figure 3.23 Crystal structure of one-dimensional arrays of inclusion complexes between **3.24** (structure shown in Figure 3.22) and C$_{60}$. PF$_6^-$ counteranions and H atoms not shown for clarity. Crystal structure refcode: CSD-ROZVOT, first published: J. C. Barnes *et. al., J. Am. Chem. Soc.*, 2015, **137**, 2392.

resorcin[4]arene ring, whose cavity depths are extended by a second layer of (usually) aromatic rings. This is achieved either by adding to the 2-position of the resorcinol ring (**3.25** in Figure 3.24A) or by bridging the adjacent resorcinol hydroxyl groups with appropriate groups (**3.26** in Figure 3.24B).

In 2005, the first water-soluble octa-acid cavitand **3.26**, whose hydrophobic deep bowl-shaped cavity facilitates the strong binding of hydrophobic guests, was reported. For example, adamantane carboxylate is bound with a K$_a$ value in water, more than an order of magnitude larger than that for the similarly sized β-cyclodextrin in water. The concave U-shaped cavity of such cavitands can also induce flexible guests to adopt unnatural conformations when compressed within a confined space. Cavitand **3.27** (Figure 3.25), for instance, forces n-alcohols (n > 10) to adopt a J-shaped conformation within the cavity, anchored by the polar hydroxyl group which remains hydrated at the mouth of the cavitand. The same reason accounts for long-chain α,ω-diols or diamines adopting a U-shaped conformation which can be exploited for macrocyclization re-actions. Using this approach, 17- to 25-membered dilactams could be synthesized in excellent yields using cavitand **3.28** (Figure 3.26, Table 3.1).

When larger hydrophobic guest molecules are present, these cavitands are also able to self-assemble in a supramolecular head-to-head fashion to form a capsule which traps the guest within the resulting cavity. Examples of these are discussed in Chapter 4. These molecular capsules have proven useful for separating gaseous hy-drocarbons, and can be used to influence the outcome of reactions of guests trapped within the cavities.

Carcerands and hemicarcerands

Developed by Nobel laureate Donald Cram in the mid-1980s, carcerands are molec-ular containers with concave internal cavities capable of encapsulating one or more guest molecules. These molecular containers, which are closed at both ends, can be thought of as two cavitand 'bowls' covalently joined together in a 'head-to-head' manner (Figure 3.27). The term 'carcerand' is a derivative of the word 'carcer' (Latin), meaning prison. Indeed, when guest molecules are imprisoned within carcerands to form '**carceplexes**', often during the synthesis process itself, they are perma-nently held and cannot be released from the molecular container even at elevated

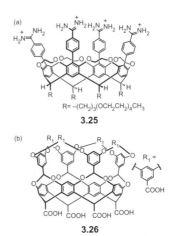

Figure 3.24 Structures of deep-cavity cavitands based on resorcin[4]arenes.

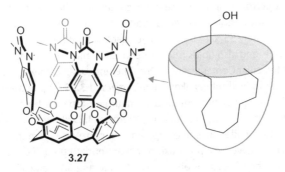

Figure 3.25 J-shaped conformation of a C_{14}-alcohol within the hydrophobic cavity of cavitand **3.27**.

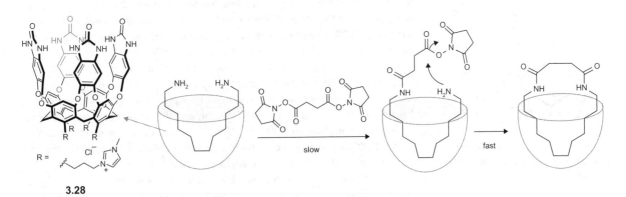

Figure 3.26 Synthesis of large dilactam macrocycles from long-chain α,ω-diamines facilitated by cavitand **3.28**.

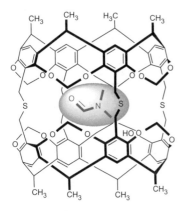

Table 3.1 Percentage yields of dilactam macrocycles synthesized from aliphatic diamines with and without cavitand **3.28**. Reprinted with permission from Q. Shi *et. al., J. Am. Chem. Soc.,* 2016, **138**, 10846. Copyright (2016) American Chemical Society.

Diamine	With cavitand 3.28	Without cavitand	Factor of enhancement
C_{11}	87	11	7.9
C_{12}	84	13	6.5
C_{14}	90	9	10.0
C_{16}	68	12	5.7
C_{18}	54	10	5.4

Figure 3.27 The first carceplex synthesized by Donald Cram containing an imprisoned molecule of DMF within the molecular container.

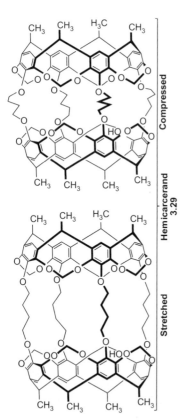

Figure 3.28 Hemicarcerand **3.29** with flexible side-arms capable of stretching to accommodate guests of different sizes.

temperatures without physical destruction of the carcerand host. This is due to steric reasons, as the openings on the carcerand are too small to allow the guest molecules to pass through.

To allow the cavities of these molecular containers to become more accessible to guest molecules, Cram and co-workers then modified the structures of the carcerands where either one or more of the spacer groups covalently linking both 'bowls' are larger, or absent altogether. The resulting larger entry and exit portals allow slow guest release or exchange at elevated temperatures. These modified molecules are called **hemicarcerands**, and likewise, their complexes with guests are called *hemicarceplexes*. The guest molecules which can form the most stable complexes tend to occupy approximately 55 % of the available space within the cavity—this is known as Rebek's 55 % rule, and also holds true for deep cavity molecular containers which are discussed in Molecular Containers with One Opening: Deep Cavity Cavitands earlier in this section. Due to its flexible linkers, hemicarcerand **3.29** (Figure 3.28) forms stable complexes with a range of different sized aromatic guest species, benzene, para-xylene and 1,4-diiodobenzene. The internal cavity volume of **3.29** almost doubles in the hemicarceplex with 1,4-diiodobenzene compared with benzene, where the host molecule is stretched along its vertical polar axis to accommodate the larger guest. For these guest molecules, the 55 % rule is obeyed. Hemicarcerand **3.29** binds 1,4-disubstituted benzene guests with greater stability than 1,2- or 1,3-disubstituted analogues due to the ease with which the host can elongate in the vertical direction. This preference allows **3.29** to be used for the separation of xylene isomers: heating **3.29** in 98% HPLC grade ortho-xylene (with the other 2% comprising of meta- and para-xylene) for four days resulted in the **3.29**-p-xylene complex being the dominant hemicarceplex despite this isomer's much smaller abundance in the HPLC solvent mixture.

When trapped inside the molecular cavities of hemicarcerands, guest molecules experience a very different environment compared to the bulk phase or in solution. For instance, the guest is now isolated within a non-polar hydrophobic environment and is no longer solvated by the bulk solvent, and the rigidity of the cavity limits guest molecular motion and conformational changes. Chemical reactions can be performed within these cavities, where the hemicarcerand now assumes the role of a molecular reaction flask. In recent years, this has led to some very exciting discoveries, such as changing the regioselectivity and rates of reactions, or allowing the isolation and observation of highly reactive fleeting intermediates. The 'taming of

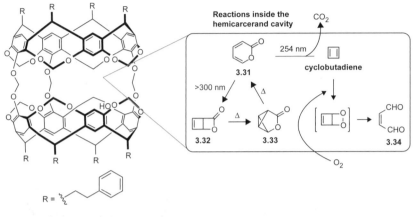

Hemicarcerand 3.30

Figure 3.29 Donald Cram's hemicarcerand **3.30** and the photochemical reactions occurring inside the cavity which converts α-pyrone **3.31** to cyclobutadiene. (D. J. Cram *et. al.*, *Angew. Chem. Int. Ed. Engl.*, 1991, **30**, 1024)

cyclobutadiene' is a good example of the latter. Cyclobutadiene is a highly reactive, severely bond angle-strained antiaromatic hydrocarbon with 4 π-electrons. By first trapping a molecule of α-pyrone **3.31** inside hemicarcerand **3.30** (Figure 3.29), a series of photochemical reactions within the molecular cavity converted **3.31** to photopyrone **3.32** which rearranged to **3.33** when heated to 90 °C in the solid state. When heated at a higher temperature of 140 °C for 12 hours, **3.33** converts back to **3.31**. However, when hemicarceplex **3.30-3.31** was irradiated with unfiltered UV-light (254 nm), photolysis of **3.31** occurs to form cyclobutadiene quantitatively, expelling a molecule of CO_2. The cyclobutadiene isolated within the molecular cavity remained stable for its ^{1}H NMR spectrum to be recorded, whose sharp and well-resolved resonances indicated rapid rotation within the cavity on the NMR timescale. When O_2 was vigorously bubbled into the solution of the **3.30**-cyclobutadiene hemicarceplex at 25 °C, the gas was able to enter the molecular cavity, converting cyclobutadiene to malealdehyde **3.34**.

3.4 Recognition of neutral guest molecules in solids

As we have seen, the field of developing artificial host molecules to selectively recognize neutral guest molecules in solution has matured rapidly. More recently, there has been increasing attention to use porous solids for selective trapping, removal, and sensing of molecules, particularly in the gas phase. Other than their important use to detect and trap toxic, dangerous gaseous molecules (e.g. chemical warfare agents and explosives) for removal, some of these porous solids can be engineered to remove CO_2 from industrial flue gases or even from air for carbon capture and sequestration. A detailed survey of this burgeoning field is beyond the scope of this primer, but the interested student can find comprehensive reviews in Section 3.6 (Further reading).

(a)   (b)

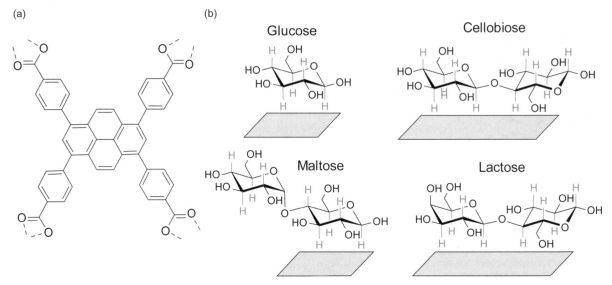

Figure 3.30 Molecular origin of carbohydrate selectivity within the NU-1000 MOF: (A) pyrene spacer units in the MOF; (B) adsorption of carbohydrates onto the pyrene spacer units. The grey panels indicate the likely area of adsorption onto the pyrene surfaces.

To impart selectivity, the pore sizes of the solid can be engineered such that only the desired target can enter the pores. Self-assembled highly porous three-dimensional crystalline solids containing ligands rigidly held in place by metal cations, known as metal–organic frameworks (MOFs) achieve this pore size control by changing the length and functional groups on the organic spacers. For instance, the presence of amines, phosphonates, and sulfonates on the organic ligand can facilitate CO_2 capture preferentially over N_2. The self-assembly and properties of MOFs are discussed in greater detail in Chapter 4.

Engineering selectivity for larger neutral molecules in MOFs can be quite challenging, as the guest preference in the solid state depends significantly on how energetically favourable the adsorption of the molecule is within the rigid pores. Nonetheless, the NU-1000 MOF with a pyrene-containing organic bridge between a Zr_6-polyoxometallate cluster selectively adsorbs disaccharides cellobiose and lactose from aqueous solution over maltose and glucose (Figure 3.30). The origin of selectivity occurs from the CH···π interactions between the carbohydrate and the aromatic planar surface of the pyrene spacer units. The β-glycosidic bond of cellobiose and lactose allows both monosaccharide units on each molecule to interact with the pyrene π-aromatic surface. In contrast, the α-glycosidic bond of maltose allows only one of the two monosaccharide units to interact with pyrene at any one time, such as in the case for glucose.

Other than MOFs, porous crystals of purely organic molecules, known as **covalent organic frameworks** (COFs) have also demonstrated guest recognition. As shown in Figure 3.31, the structurally rigid macrocycle **3.35** was able to form a porous crystalline highly fluorescent structure, due to the tetraphenylethylene motifs*. When air saturated with trinitrotoluene (TNT) vapour was bubbled into a suspension of these empty porous

* The tetraphenylethylene motif is a popular molecular unit displaying aggregation-induced emission (AIE), where aggregation of these units enhances the fluorescence intensity. This occurs as aggregation reduces the molecular flexibility, concomitantly minimizing non-radiative pathways for energy loss upon photoexcitation.

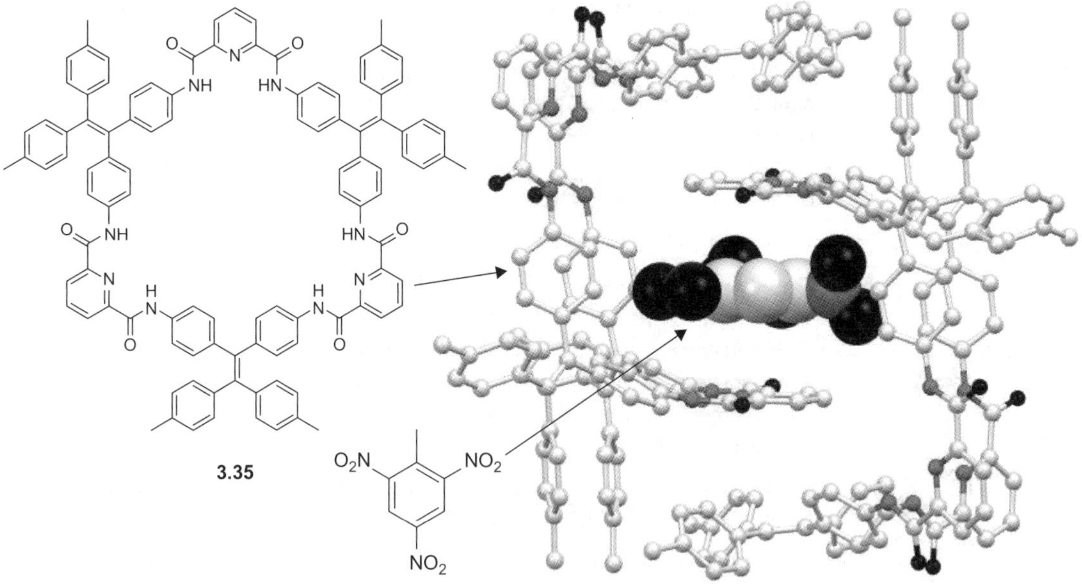

Figure 3.31 TNT adsorption within the porous crystal structure of macrocycle **3.35**. Crystal structure refcode: CSD-MESGOJ, first published: J. B. Xiong *et. al.*, *Org. Lett.*, 2018, **20**, 321.

crystals in THF/ H_2O, TNT was adsorbed within the crystalline structure. This causes fluorescence quenching, which was used to detect TNT with femtogram sensitivity. Analysis of the crystal structure showed that the electron-deficient TNT aromatic ring π-stacked with the pyridyl ring of each macrocycle **3.35**. In addition, highly unusual strong interactions between the lone electron pairs on the nitro-groups of TNT and the pyridyl ring were found (termed $n\cdots\pi$ interactions, where n referred to the lone pairs), which contributed significantly to TNT binding and adsorption within the crystal. Other than TNT, these porous crystals of **3.35** could also adsorb CO_2 selectively over N_2.

3.5 Summary

Although the selective binding of neutral guests is amongst the most difficult challenges in molecular recognition, in the last decade, tremendous strides have been made. We have seen how a judicious analysis and understanding of the guest structure can be used to generate highly complementary hosts exhibiting exemplary selectivity, such as the caged glucose receptor shown in Figure 3.4. Other modern approaches, including the use of database-driven computer-aided molecular host design, and an artificial 'molecular nose', also show great promise. Macrocyclic hosts containing hydrophobic, solvent-shielded inner cavities are of great importance in binding small organic molecules, especially in aqueous solvents driven by the hydrophobic effect. Together, the binding and sensing of neutral guest molecules have led to important industrial applications, including glucose monitoring for diabetic patients, deodorization and stabilization of unstable bioactive molecules in the consumer care industry, nanotechnology and smart materials development, just to name a few. We can expect that the coming years will witness new ingenious supramolecular approaches to expand the library of neutral molecular targets, driving new technological developments.

3.6 Further reading

As for cation and anion guest species in Chapter 2, the rapid developments in the field of neutral guest recognition have witnessed explosive growth in the last two decades. This chapter can only provide a brief glimpse into the ingenuity and exciting research activity taking place worldwide. The following articles provide more in-depth discussion of some of the diverse areas of host–guest recognition we have considered in this chapter:

- Supramolecular receptors for glucose binding: *Chem. Rev.*, 2015, **115**, 8001–37.
- Aqueous supramolecular chemistry of cucurbit[n]urils, pillar[n]arenes, and cavitands: *Chem. Soc. Rev.*, 2017, **46**, 2479–96.
- Harada's own account of his seminal contributions towards the supramolecular chemistry of cyclodextrins: Chapter 10 of Izatt, R. M. (Ed.), 2016, *Macrocyclic and Supramolecular Chemistry: How Izatt–Christensen Award Winners Shaped the Field*, Chichester: Wiley.
- The use of optical sensing approaches towards supramolecular analytical chemistry: *Chem. Rev.*, 2015, **115**, 7840–92
- The formulation of Rebek's 55 % rule for dimensions of guest binding inside hydrophobic cavities: *Chem. Eur. J.*, 1998, **4**, 1016–22.
- Supramolecular chemistry in solid materials such as MOFs: *Isr. J. Chem.*, 2018, **58**, 1102.

3.7 Exercises

3.1   Boronic acids have been exploited for chiral discrimination of α-hydroxycarboxylates using the indicator displacement assay (IDA) approach. As shown in Figure 3.32, binding of the chiral substrate displaces the dye (pyrocatechol violet, or **PV**), which is originally coordinated to the boronic acid receptor. This leads to an enhancement in fluorescence intensity.

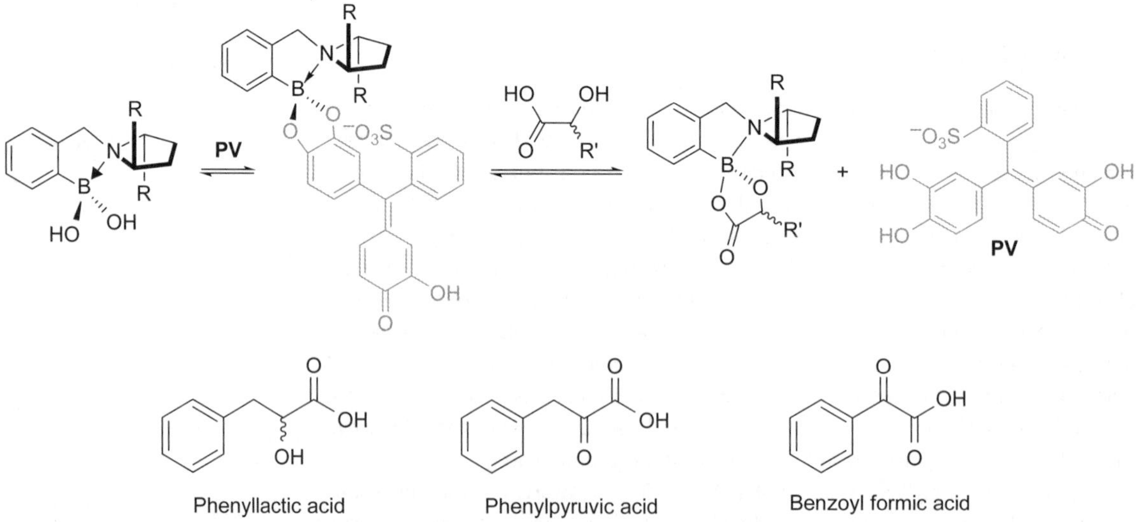

Figure 3.32 Binding of PV to a boronic acid receptor and its use in IDAs for the chiral discrimination of α-hydroxycarboxylates.

(a) Explain how the chiral boronic acid receptor (R = CH$_2$OCH$_3$) is able to discriminate between chiral hydroxycarboxylates.

(b) When R = CH$_2$OCH$_3$, the enantioselectivity (K_L/ K_D) for phenyllactic acid is 2.82. However, the selectivity diminishes to 1.48 when R = CH$_3$. Suggest a possible reason for this observation.

(c) Despite not possessing a diol functionality, phenylpyruvic acid binds to the achiral version of the receptor (i.e. when R = H) almost as strongly as the dye pyrocatechol violet. However, the structurally similar benzoyl formic acid showed very weak binding to the same boronic acid host. Explain this difference.

3.2 A series of compounds are shown in Figure 3.33 which are able to self-dimerize in solution:

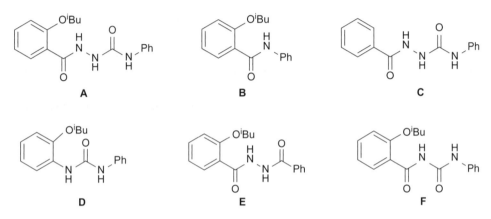

Figure 3.33 Structures of molecules capable of self-dimerization by hydrogen bonding.

(a) Draw the structures of the hydrogen bonded dimers for these molecules.

(b) Arrange these molecules in increasing dimerization constants, starting with the weakest. Justify your choice.

(c) For compound A, ^{1}H NMR analysis of the dimer in CDCl$_3$ showed three distinct chemical shifts for the N–H protons engaged in hydrogen bonding at 7.4, 8.3, and 10.4 ppm. Suggest and justify a likely assignment of these NMR resonances.

3.3 The abilities of macrocycles such as cyclodextrins to form inclusion complexes with hydrophobic molecules in aqueous solvents have suggested to chemists the possibility of incorporating them into catalyst design to perform reactions in water. One such example is Ronald Breslow's bis-βCD catalyst **3.36**, shown in Figure 3.34, bridged by a Cu(II)-coordinating 2,2-bipyridyl unit. This molecule catalyses the hydrolysis of hydrophobic esters (such as **3.37**) in water, accelerating the reaction more than 18,000 times at pH 7 and 310 K.

(a) Suggest a plausible catalytic mechanism for the hydrolysis of substrate **3.37**, drawing a catalytic cycle to support your answer.

(b) The activities of some catalysts can be diminished by strong product–catalyst binding inhibition, decreasing the ability of the original substrate to bind to the catalyst. Explain how the structure of **3.36** prevents such 'catalyst poisoning' from occurring.

(c) How would you expect the rate of ester hydrolysis catalysed by **3.36** to change when the reaction takes place in ethanol compared to water?

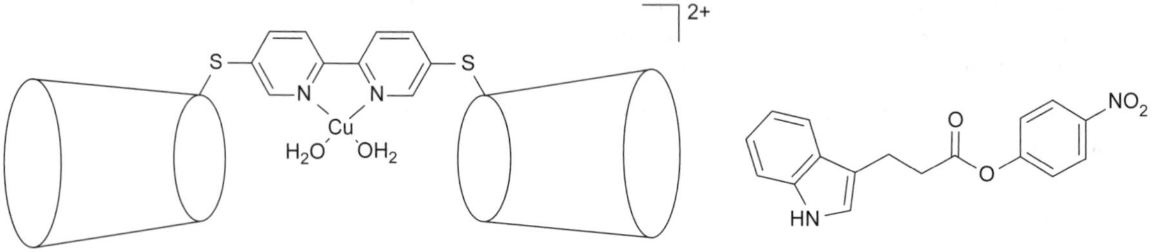

3.36   **3.37**

Figure 3.34 Breslow's bis-cyclodextrin catalyst **3.36** and a hydrophobic ester substrate **3.37**.

For the answers to these exercises, visit the online resources which accompany this primer.

4 Self-assembly

4.1 Introduction

The previous chapters introduced the concept of molecular recognition, the process in which a host molecule binds a cationic, anionic or neutral guest species to form a non-covalent host–guest complex. In the majority of cases, these complexes contain only two, sometimes three, components. However, in this chapter, we describe the formation of larger supramolecular assemblies for which this number will often be much higher. As such, the distinction between host and guest will become blurred and instead, the molecular components that associate to form these larger and more complex supramolecular assemblies ('supermolecules') are referred to as building blocks (Scheme 4.1). Crucially, the intrinsic information contained within these molecular building blocks (size, shape, electronic characteristics) dictate the nature and orientation of their intermolecular interactions with one another (hydrogen bonding or aromatic stacking) and subsequently, their global arrangement into a specific superstructure. This process, involving multiple recognition events between molecules or ions, is reversible, spontaneous, and termed **self-assembly**.

Self-assembly is prevalent within Nature. Cell membranes, multi-component enzymes, and the iconic DNA double helix all owe their unique three-dimensional structures to biomolecules that assemble in a specific manner via combinations of non-covalent interactions. In the latter case, the efficacious self-assembly of two strands of deoxyribonucleic acid is driven by hydrogen bonding, hydrophobic forces, and π–π stacking interactions (Figure 4.1). Selectivity between complementary guanine–cytosine (G–C) and adenine–thymine (A–T) base pairs makes this an example of strict self-assembly.

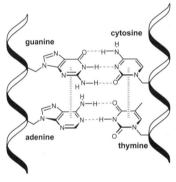

Figure 4.1 Self-assembly of the DNA double helix uses both complementary hydrogen bonding and π–π stacking interactions. This is shown here as Watson–Crick base pairs of guanine-cytosine (G–C) and adenine–thymine (A–T).

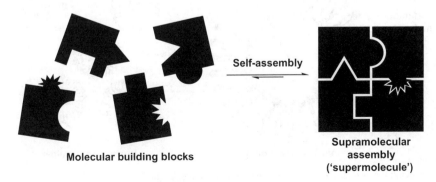

Scheme 4.1 A simplified illustration of molecular self-assembly.

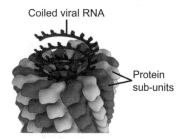

Coiled viral RNA

Protein sub-units

Figure 4.2 A cartoon of the self-assembled tobacco mosaic virus. Adapted under the terms of the Creative Commons Attribution License from Wikipedia.

Viruses are infectious agents that provide a particularly eloquent and dangerously effective demonstration of the power of self-assembly. In general, a virus consists of genetic material (strand(s) of RNA or DNA) surrounded by a protective coat of protein (capsid) that, in the case of the tobacco mosaic virus shown in Figure 4.2, is formed by the self-assembly of 2130 protein sub-units, generating the viral superstructure. Viruses such as influenza have an additional lipid bilayer component to their outer protein coat, often adopted from the cell membrane of the infected host cell and known as a viral envelope. This envelope is also an example of self-assembly; amphiphilic molecules with hydrophobic tails (fatty acid chains) and hydrophilic heads (phosphate groups) orientate themselves into a bilayer structure so as to maximize favourable intermolecular interactions with each other and with water molecules of the surrounding medium (Scheme 4.2). Whilst this robust structure can make the virus deadly inside the body, supramolecular chemistry provides a way to destroy it before it can get there. Surfactant molecules (e.g. soaps) are also **amphiphiles**, and on coming into contact with the virus will assemble themselves into the bilayer, fracturing it and breaking it open because of their different shape and size. This causes the virus to be disassembled and, in fact, permanently disabled via the trapping of its constituent parts within self-assembled micelles of the surfactant (Scheme 4.2).

In the laboratory, self-assembly has allowed chemists to construct supramolecular ensembles of beauty and complexity that are inaccessible using traditional multi-step covalent bond-making and-breaking syntheses. These architectures can reach the size of proteins and are produced in high yield (often quantitative) by the mixing of simpler, yet appropriately designed, sub-units in a one-pot synthesis. This is because synthesis by self-assembly is performed under thermodynamic control. By careful design, discrete supramolecular assemblies are formed as the **thermodynamic product** as the system strives to the energetic minimum through an interplay of attractive and repulsive interactions (enthalpy) while minimizing the loss of degrees of freedom (entropy). The relevance of this research to biology is apparent from the above; however, new applications are being realized from the construction of abiotic molecular assemblies as functional chemical transporters and catalytic nanoreactors.

This chapter is arranged so as to present cation, neutral, and anion self-assembly in separate sections. However, there are, of course, parallels between them all, not least

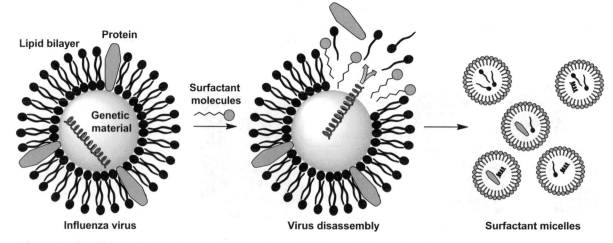

Scheme 4.2 The self-assembled structure of an influenza virus and its disassembly by a surfactant, e.g. soap.

the non-covalent interactions used to drive or template the process. We have also tried to highlight that self-assembly is not just a tool used in molecular synthesis. It is successfully employed across a range of length scales from the generation of discrete molecular cages to infinite porous framework materials, with some further applications of these supramolecular materials detailed towards the end of this chapter.

4.2 Metal cation self-assembly

The classical metal cation template effect used for the synthesis of macrocyclic ligands was pioneered by Daryle Busch. The preorganization of reactants within the coordination sphere of a metal ion-directed cyclization reaction disfavours the formation of oligomeric side products. This metal template effect methodology was expanded upon by exploiting the stereochemical preferences of metal ion(s) to dictate the organization of *multiple* polydentate organic ligands into larger supramolecular architectures. In these the metal ion coordinate bond is an integral component of the assembled polymetallic structure. This form of self-assembly, termed metal-directed self-assembly, is distinct from the classical metal templated syntheses of macrocycles, cryptands, and, in particular Nobel laureate Jean-Pierre Sauvage's mechanically interlocked molecules (e.g. catenanes and rotaxanes) discussed in Chapter 5.

The range of assembled polymetallic supramolecular systems described in this section can be generally classified as either coordination complexes or coordination polymers. Importantly, supramolecular coordination complexes are discrete molecular entities, e.g. molecular cages, typically studied in the solution phase, whilst polymers are composed of infinite arrays of metal cation–ligand bonding to give a solid-state material, e.g. metal–organic frameworks. However, common to both structures are the creation of well-defined cavities within the assembly that are often used for recognition of guest molecules.

The majority of this section focuses on *transition metal*-directed self-assembly where metal–ligand coordinate (dative covalent) bond strength and kinetic lability can be tuned by ligand variation and choice of metal. One can imagine that such properties are highly desirable for the uptake, transport, and release of guest molecules, e.g. drugs, from the cavities of these assemblies. A further advantage of this approach is to exploit highly directional metal ligand field effects to dictate the shape of the assembled transition metal polymetallic complexed structure e.g. tetrahedral vs. square planar geometries for four coordinate systems. Hence, by using transition metals as templates in combination with a specific polydentate ligand design, the shape and size of the resulting assembled polymetallic host cavity can potentially be tuned to recognize a target guest substrate. Finally, redox- and photo-active transition metals can impart sensing and catalytic properties in such assemblies.

Edge-directed approach

Combining carefully designed multi-dentate organic ligands with coordinatively unsaturated transition metal complexes in a strict stoichiometric ratio can result in an enormous variety of polymetallic architectures, ranging from macrocycles to cages and even polyhedral spheres of nanometre dimensions. The stereochemical preference of the metal (on the vertices) and polydentate ligand design (on the edges) dictates the symmetry and shape of the final polymetallic assembled product, often forming in high yield.

One of the earliest demonstrations of transition metal-directed assembly is molecular square **4.1**, formed from the self-assembly of eight components; four 'corner' metal complexes connected with four rigid linear ligand edge spacer units (Scheme 4.3). Critical to the formation of this macrocycle is the square planar coordination geometry preference of the $4d^8$ Pd(II) metal centres which create the 90° angle required for the corners of the square. A chelating cis-bidentate ethylene diamine

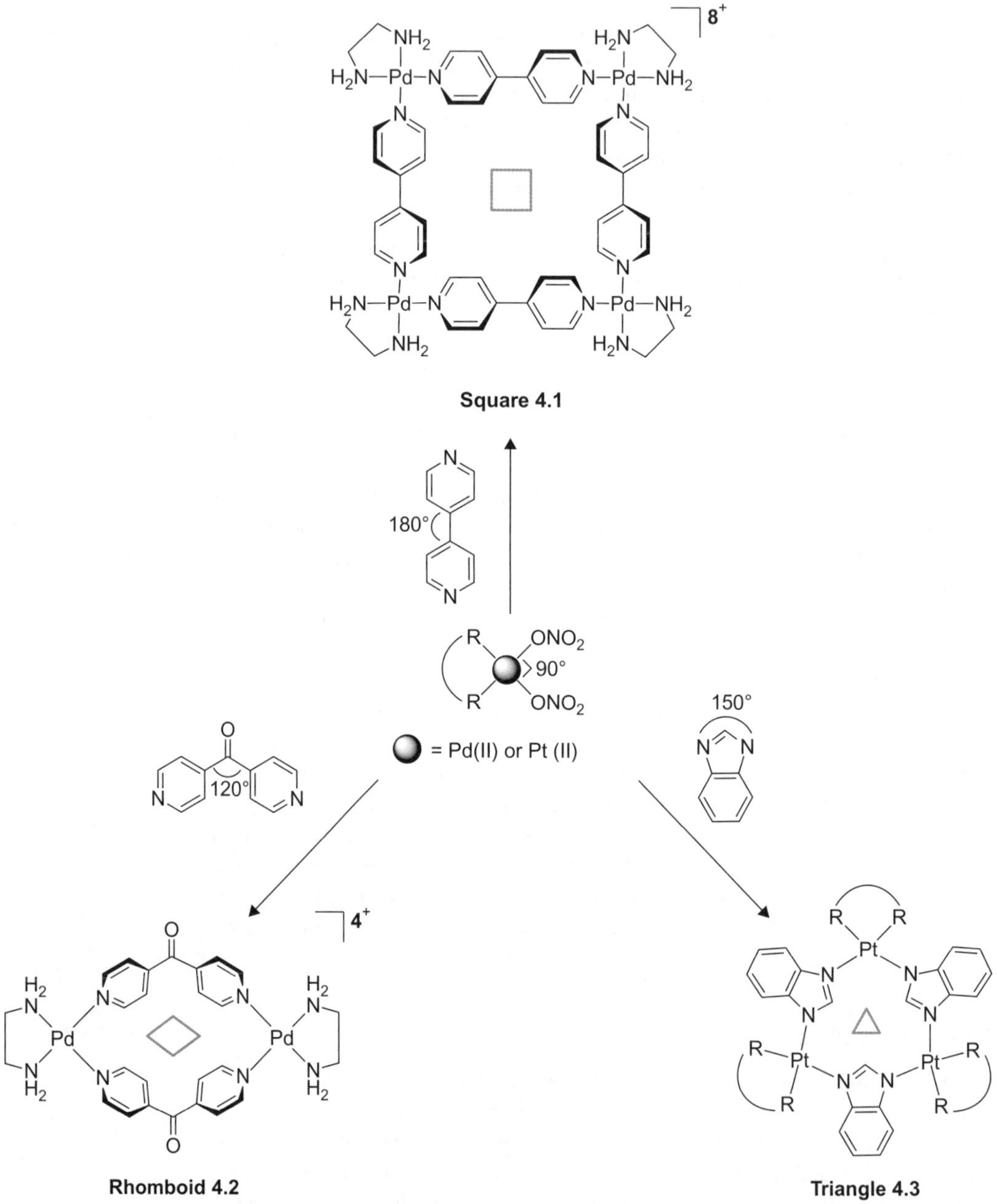

Square 4.1

Rhomboid 4.2

Triangle 4.3

Scheme 4.3 Macrocycles of various shapes are constructed through the self-assembly of square planar Pd(II)/Pt(II) metal complexes with bidentate organic ligands of different geometries. R = bidentate supporting ligands for Pt (II).

ligand acts as a protecting group for the two remaining coordination sites. Introduced in Chapter 1, the chelate effect ensures these ethylene diamine ligands are less labile than their monodentate nitrate anion counterparts.

Varying the nature of the edge-ligating group dictates the shape of the product. For example, altering the preorganized bite angle of the bidentate ligands from 180° (for the square) in combination with the same square planar Pd(II)/Pt(II) metal complexes, generates either a molecular rhomboid **4.2** (120°) or molecular triangle **4.3** (150°) assembly (Scheme 4.3). Notably, the design of a Pd(II) complex with a single vacant coordination site enabled the synthesis of a hexanuclear 'palladawheel' assembly **4.4** in high yield (Figure 4.3). In general, these macrocyclic products are thermodynamically favoured over oligomeric and polymeric side products since they have fewer monomer units (reducing entropic cost) while maximizing bonding interactions (enthalpic gain).

The two-dimensional molecular square **4.5** contains long, linear, rigid ligands incorporating a perylene diimide dye molecule, and is one of the largest of its kind with a diagonal length of 3.4 nm (Figure 4.4). Interestingly, the fluorophores of square **4.5** are strongly emissive because of their large separation from the metal corners,

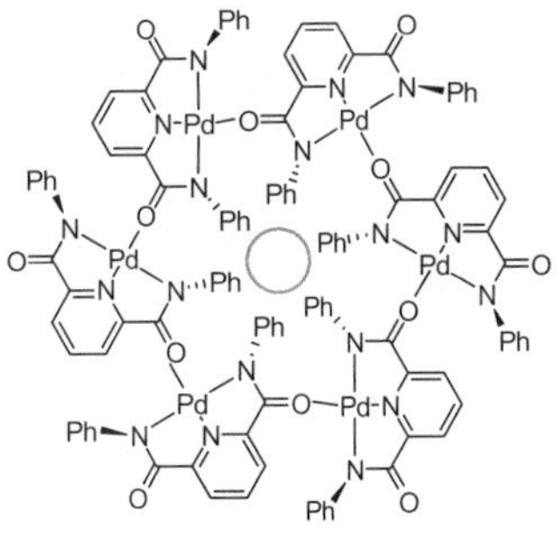

Wheel 4.4

Figure 4.3 A hexanuclear Pd(II) wheel assembly with phenyl substituents acting as interior 'spokes' and exterior 'propellers'.

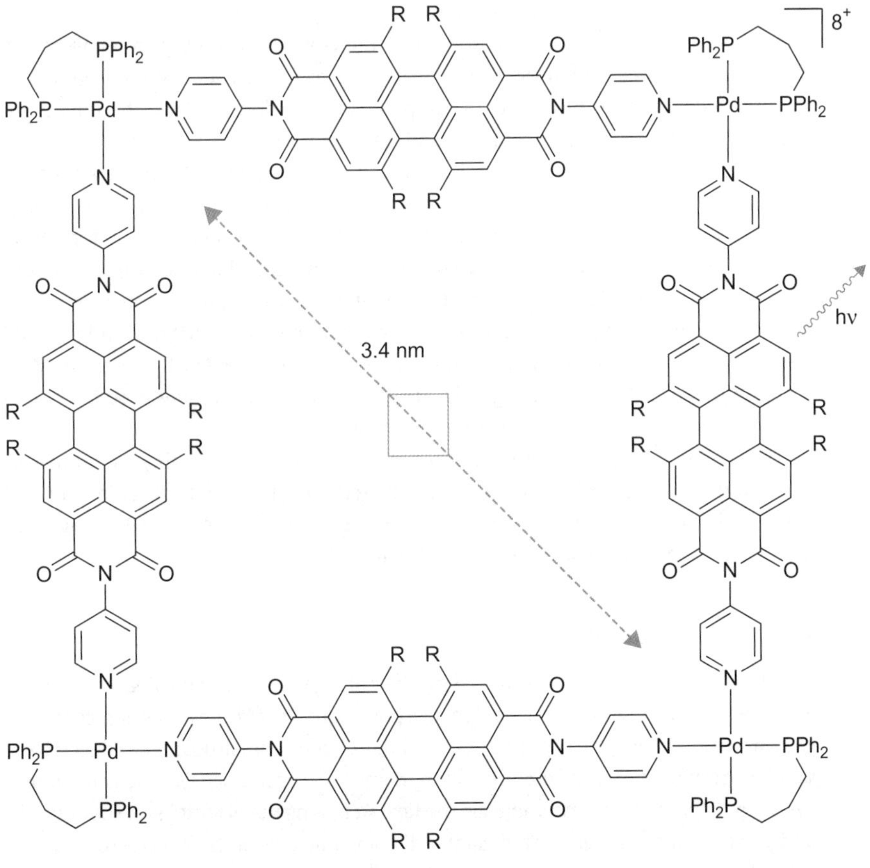

Square 4.5

Figure 4.4 A molecular square assembled with fluorophore-based ligands along the edges and Pd(II) complexes at the corners. R = solubilising groups

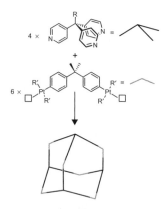

Adamantoid cage 4.7

Figure 4.6 A set of metal complexes and ligands containing pre-programmed tetrahedral bond angles (109.5°) are used to direct the self-assembly of an adamantoid cage.

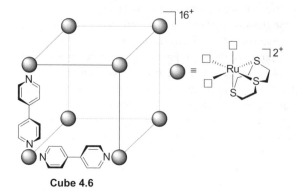

Cube 4.6

Figure 4.5 The self-assembly of twelve bi-pyridyl ligands and eight octahedral Ru(II) complexes, each with three labile coordination sites, generates a molecular cube. Boxes represent vacant coordination sites on the metal complex.

opening up the possibility of optical sensing of guest molecules such as polyaromatic hydrocarbon pollutants encapsulated within the cavity.

Beyond two dimensions, transition metal-directed self-assembly has enabled the construction of three-dimensional ensembles, such as molecular cube **4.6** (Figure 4.5). The cube is comprised of twelve 4,4′-bi-pyridyl ligand donors along the edges and eight Ru(II) complexes at the corners and was synthesized by heating a mixture of these components for one month. Long reaction times and high temperatures are typical for forming complex cage-like assemblies employing kinetically inert metal ions (i.e. Ru(II) is low spin $4d^6$ electron configuration). This is because the formation of the thermodynamic product (cube **4.6**) is a slow process due to competition from relatively stable, kinetically trapped intermediates. Over time, the gradual formation of the symmetric cube product species (from twenty self-assembling molecules) could be followed by the simplification of the reactant mixture's ^1H NMR spectrum. Mass spectrometry and X-ray crystallography are also common techniques used to characterize large supramolecular coordination complexes.

The synthetic possibilities arising from metal-directed self-assembly are numerous and reports of various polyhedra can be uncovered in the Further reading (Section 4.7) at the end of this chapter. One particularly exotic example is assembly **4.7** which has the shape of a molecule of adamantane (Figure 4.6).

Face-directed approach

Whilst in an *edge-directed* synthesis the ligands lie along the edges of the final assembly, ligands can also be positioned on the faces of a polyhedron, the *face-directed* approach, also known as molecular panelling. The desired three-dimensional structure is reduced to its simpler constituent molecular components, namely the ligand panels which are synthesized separately and then assembled in the final step using metal templates. For example, a trigonal prism can be generated via the stitching together of three molecular square-shaped panels. As shown in Scheme 4.4, these ligand panels (L) are comprised of a porphyrin unit that has four pyridyl donor groups to enable its coordination with the familiar square planar cis-ethylene diamine Pd(II) metal complex (M), except now the metals are positioned along the edges. A 2:1 M:L ratio generates the M_6L_3 trigonal prism **4.8**, whilst a

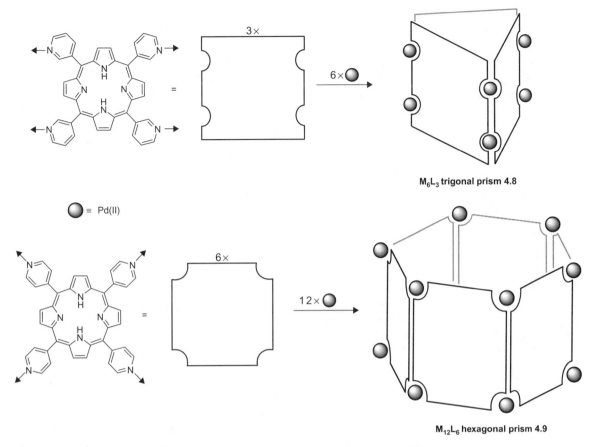

M_6L_3 trigonal prism 4.8

◯ ≡ Pd(II)

$M_{12}L_6$ hexagonal prism 4.9

Scheme 4.4 The face-directed approach provides an alternative route to the self-assembly of three-dimensional architectures such as trigonal and hexagonal prisms.

subtle change in the substitution pattern of the pyridyl donor forms the larger $M_{12}L_6$ hexagonal prism **4.9**. When mixtures of different sized panels are used then many caged products could result. However, it is possible to control the product outcome, e.g. favour the formation of the larger or smaller cage, by using a molecular guest of complementary size or shape to that of the desired cage to act as a further template during the synthesis. This will be discussed later in this chapter in the section on additional templating interactions.

Symmetry interaction approach

The above examples used transition metal complexes with partially vacant coordination sites. However, it is possible to exploit free transition metal ions with only weakly coordinating counteranions as building blocks for self-assembly. This strategy, named the symmetry interaction approach, relies on the inherent geometry of the interaction between a free metal ion and a chelating ligand.

Early examples include molecular grids and **helicates** from the group of Nobel laureate Jean-Marie Lehn. The former involves mixing multidentate linear pyridine and pyridazine-based ligands with free Ag(I) ions (the counterions are non-coordinating triflate anions) in a 2:3 molar ratio to form a 3 × 3 molecular grid **4.10** (Scheme 4.5). In

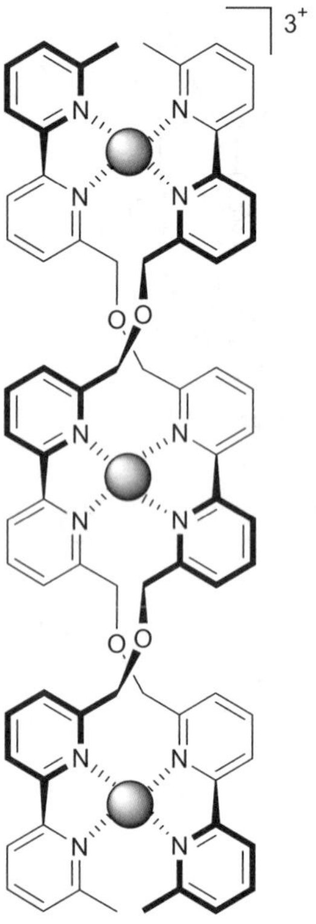

M₃L₂ helicate 4.11

○ ≡ Cu(I)

Figure 4.7 The tetrahedral geometry of four-coordinate Cu(I) ions enables self-assembly of a metal–ligand double helix.

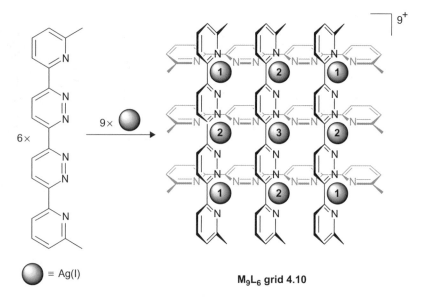

○ ≡ Ag(I)

M₉L₆ grid 4.10

Scheme 4.5 The self-assembly of a 3 × 3 molecular grid using the symmetry interaction approach. The number labels on the Ag(I) ions refer to their different environments as observed by ^{109}Ag NMR spectroscopy.

total this involves the self-assembly of 15 discrete molecular components. The structure was confirmed using ^{109}Ag NMR spectroscopy which shows the three distinct Ag environments in a 4:4:1 integration ratio.

Ligands of similar design, typically in combination with tetrahedral directing Cu(I) cations, have been used to assemble molecular helicates in which two identical multidentate ligand strands dimerize, generating structures that are reminiscent to that of the DNA double helix, where the metal cation replaces the complementary hydrogen bonding between the strands. Importantly, in comparison to the previous poly-pyridyl ligand, the one used for helicate **4.11** in Figure 4.7 contains ether linkages making it less rigid and allowing it to twist as it coordinates the three Cu(I) ions. Remarkably, when Cu(I) metal ions were added to a mixture of polybi-pyridyl ligands of differing length, only helicates containing two strands of the same length were formed. This outcome is known as self-recognition and is a key mechanism in the recognition and replication of DNA in Nature.

Akin to the previous molecular cube and prisms (Figure 4.5 and Scheme 4.4), the symmetry of metal–ligand interactions has been used in the construction of numerous three-dimensional molecular assemblies, known more generally as coordination cages. A common shape is the tetrahedral cluster with formula M₄L₆, in which the four metal ions occupy the vertices and the six ligands make up the edges of a tetrahedron as shown in Scheme 4.6. A key difference between these assemblies and previous edge-directed syntheses are that the metal ions are 'free' in solution, i.e. they are not metal complexes with inert sets of ligand sets maintained throughout the synthesis.

The four Ga(III) ions of tetrahedron **4.12** are pseudo-octahedrally coordinated by one catechol bidentate arm from each of three separate ligands. Furthermore, the naphthalene spacer incorporated into the ligand component imparts rigidity to inhibit formation of the more entropically favoured M₂L₃ helicate **4.13** (Scheme 4.6). This structure is also notable because it shows that main group ions with defined coordination geometries can be used, as discussed in more detail later in this chapter.

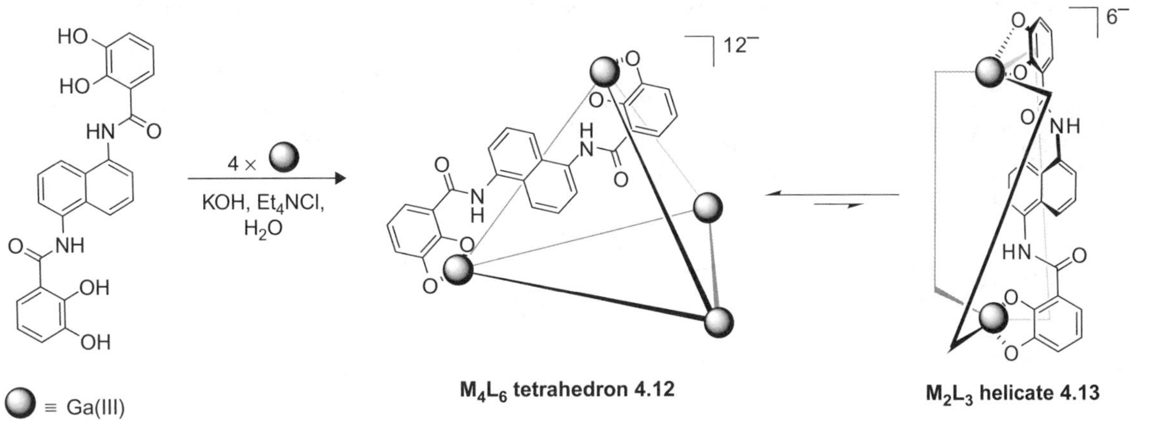

M_4L_6 tetrahedron 4.12

M_2L_3 helicate 4.13

= Ga(III)

Scheme 4.6 The self-assembly of an M_4L_6 tetrahedron is enthalpically favoured over the analogous M_2L_3 helicate because of the rigidity of each ligand strand.

An extravagant metal-directed polyhedron is $M_{24}L_{48}$ rhombicuboctahedron **4.14** in which 72 components are organized and arranged into an almost spherical assembly (Scheme 4.7). The internal void spaces of such structures are large enough to accommodate multiple molecular species, even other self-assembled structures, known as 'Russian doll assemblies'. This has also allowed these cages to be used as tiny reaction vessels in which the reactivity of individual molecules can be precisely studied and controlled at the nanoscale. For further applications of these structures please see Section 4.5 at the end of this chapter.

Additional templating interactions

As demonstrated in the previous sections, a variety of elegant molecular shapes and architectures have been constructed using metal-directed assembly. In all these cases it is the metal ions themselves that act as the primary templates; however, more detailed analysis of the final supramolecular assembly often reveals additional non-covalent interactions that are of importance to templating their synthesis.

It was discovered that the weakly coordinating counteranions of transition metal assembled cages are not always completely innocent in the self-assembly process. For example, an X-ray crystal structure of the M_4L_6 tetrahedron **4.15** revealed a persistent tetrahedral BF_4^- anion in the cavity that did not undergo exchange with any of the anions on the periphery of the cage (Figure 4.8). This indicated the BF_4^- anion acted as a secondary template during cage formation, forming hydrogen bonds to ligands that are polarized by their coordination to positively charged metal centres at the vertices. However, there is no evidence for BF_4^- anion templation in the assembly of a related M_8L_{12} cube (**4.16**), presumably because the cavity shape is not complementary to that of the tetrahedral guest (Figure 4.8). Instead, π–π aromatic stacking interactions between adjacent ligand heterocycles assist in the self-assembly process. Furthermore, the strong and selective binding of $[N(CH_2CH_3)_4]^+$ was shown to assist in forming assembly **4.12** on account of the tetrahedral cage's negative charge (Figure 4.8).

An exotic example of anion templation in metal-directed self-assembly is seen in Lehn's synthesis of the M_5L_5 circular helicate **4.17** in Figure 4.9. This structure is formed with five tris-bi-pyridyl ligand strands that wrap themselves around five Fe(II) ions and coordinate a single chloride anion in the central cavity through a combination of hydrogen bonding and electrostatic interactions. The chloride anion plays a crucial role

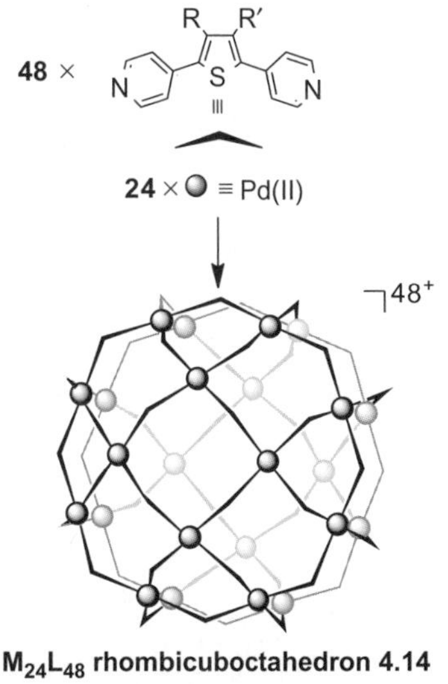

$M_{24}L_{48}$ rhombicuboctahedron 4.14

Scheme 4.7 On account of its size and complexity, the synthesis of a rhombicuboctahedron structure could only be realized via self-assembly as opposed to a step-wise approach.

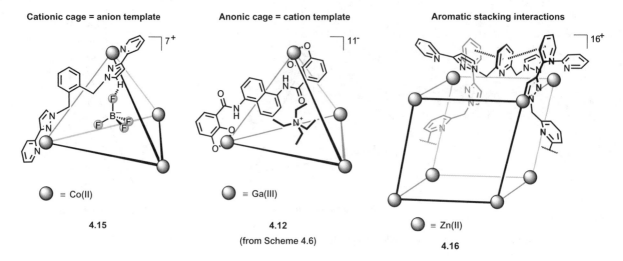

Cationic cage = anion template   **Anonic cage = cation template**   **Aromatic stacking interactions**

$\equiv$ Co(II) $\equiv$ Ga(III) $\equiv$ Zn(II)

4.15   **4.12**   **4.16**

                                                       (from Scheme 4.6)

Figure 4.8 Anion recognition provides a secondary interaction in the formation of a positively charged tetrahedral cage (4.15), while cation binding is present in an overall negatively charged assembly (4.12). Aromatic stacking interactions between ligand sets contribute to the self-assembly of a molecular cube (4.16).

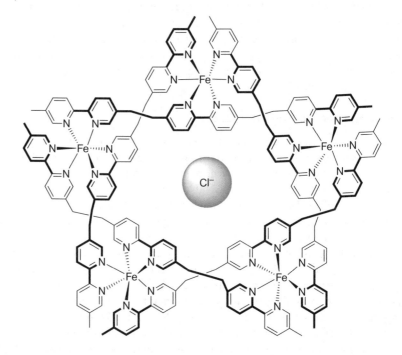

M$_5$L$_5$ circular helicate 4.17

Figure 4.9 The self-assembly of this circular helicate relies on both metal coordination and anion recognition.

in self-assembly because when the FeCl$_2$ starting material is swapped for FeSO$_4$, the larger sulfate anion drives formation of a larger M$_6$L$_6$ circular helicate. There are also examples in which the *shape* of the templating anion (e.g. tetrahedral ClO$_4^-$, trigonal planar NO$_3^-$) will influence the conformation of supramolecular architectures too.

When the overall metal coordination assembly is charge neutral and hence no competing counterions are present, then the opportunity to design a neutral molecular template to direct metal coordination during the self-assembly process can be exploited. This discrete template strategy is demonstrated during the construction of large macrocycles known as nanorings. Critical to forming nanoring **4.18** in an appreciable yield is the use of the Zn(II) metalloporphyrin–pyridyl coordination to assemble three tetra-porphyrin macrocycle precursors in a circle around a dodecapyridyl template (Scheme 4.8). This template then holds these metal-based building blocks in place during the covalent coupling of the alkyne groups. Since the dodecapyridyl ligand template coordinates twelve Zn(II) metalloporphyrin motifs, this could even be considered a ligand-directed assembly instead of a classical metal-directed one. The cavity of the nanoring is nearly 5 nm across but contracts if the template is removed from the centre. Sources in the Further reading (Section 4.7) detail a Vernier templating strategy that involves multiple smaller templates to construct even larger nanorings containing up to 24 porphyrin units.

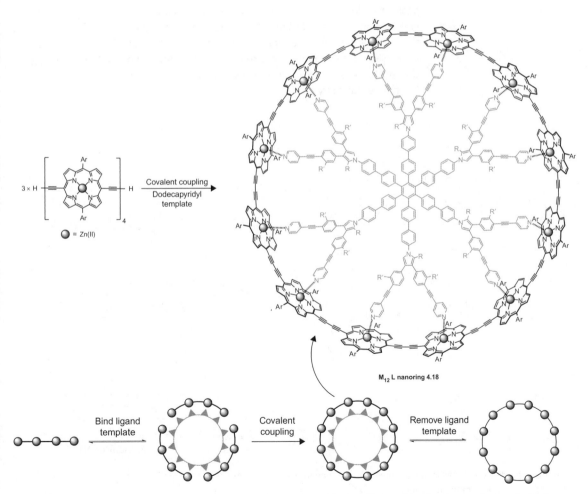

Scheme 4.8 A discrete molecular template is strategically designed to guide the self-assembly of twelve porphyrin units for the synthesis of a large macrocycle called a nanoring.

Multi metal/ligand-component architectures

Supramolecular ensembles of greater complexity and potential functionality have been produced through the coordination of more than one type of ligand to a metal ion. For example, a more rigid version of the polypyridyl ligand from helicate **4.11** (Figure 4.7) has been combined with Cu(I) cations in the presence of a second ligand type composed of two pyrimidine rings (a tetradentate donor). One might expect that a large number of new coordination networks are possible from the mixing of these three different components. However, thermodynamic control ensures the chemical system evolves towards a state in which the most binding sites available on each ligand and metal ion are filled, known as the principle of maximum occupancy. The final product in the case above is molecular ladder **4.19** in which two polypyridyl strands (rails) are connected by bis-pyrimidine ligands (rungs), a structure that satisfies the coordination number and geometric requirement of all Cu(I) metal ions and N donor atoms (Scheme 4.9).

Supramolecular architectures containing multiple metal ions are also possible and the example in Scheme 4.10 demonstrates the power of metal–ligand self-assembly to achieve this level of molecular ordering. Assembly **4.20** is a cubic array of metal ions in which octahedral Fe(II) occupies the corners and square planar Pt(II) ions cap each of the faces because the latter preferentially coordinate to the pyridyl donor groups. An impressive 62 components are assembled and 96 new bonds are formed during the one-pot synthesis, since ligand precursors are also covalently connected via reversible Schiff base chemistry.

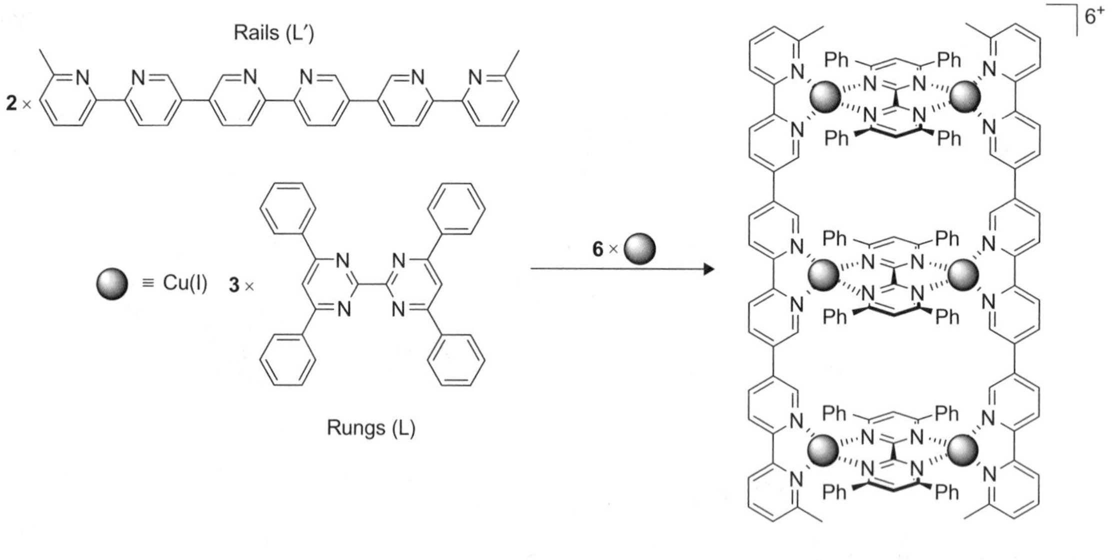

$M_6L_3L'_2$ ladder 4.19

Scheme 4.9 A self-assembled ladder architecture is formed from a three-component mixture of ligands and metal ions according to the principle of maximum occupancy.

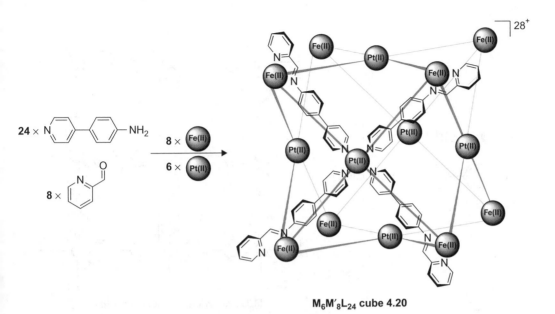

M$_6$M'$_8$L$_{24}$ cube 4.20

Scheme 4.10 A total of 62 metal and ligand components undergo self-assembly upon heating over twelve hours to generate a molecular cube in high yield. The grey lines indicate the ligands linking the metal ions (four ligands are shown on the front face).

Metal coordination polymers and frameworks

A few years after the report of metal-templated macrocycle **4.1** (Scheme 4.3), self-assembly was exploited to connect these discrete supramolecular complexes together and form an extended two-dimensional square grid (Scheme 4.11). Array **4.21** is termed a coordination polymer since, whilst the square is a finite structure and monomeric in solution, the grid consists of an infinite (polymeric) chain of Cd(II) metal complexes bridged by organic linkers which precipitates as a crystalline solid. However, just like macrocycle **4.1**, the larger assembly was able to bind guests, in this case molecules of 1,2-dibromobenzene via π–π aromatic stacking interactions, forming a material known as a clathrate.

Polymeric M–L assemblies can also be formed in three dimensions, and often the guests found within the cavities of these structures are residual solvent molecules left over from the crystallization process. For some, it is possible to remove these guests through heating while keeping the metal–ligand framework intact. Now the crystalline host material has porous channels within its structure that can be used to bind and release different guests, some of the earliest examples being gases such as methane. This property of reversible guest uptake has given rise to a class of 3D metal arrays known as metal–organic frameworks (MOFs, also see Section 3.4). The infinite array of cavities within their framework provide large internal surface areas, making them extremely porous materials and therefore attractive for applications requiring substrate adsorption, such as energy storage, chemical purification, and sensing. Although typically not as thermally stable as zeolites, the vast potential range of metal and ligand building

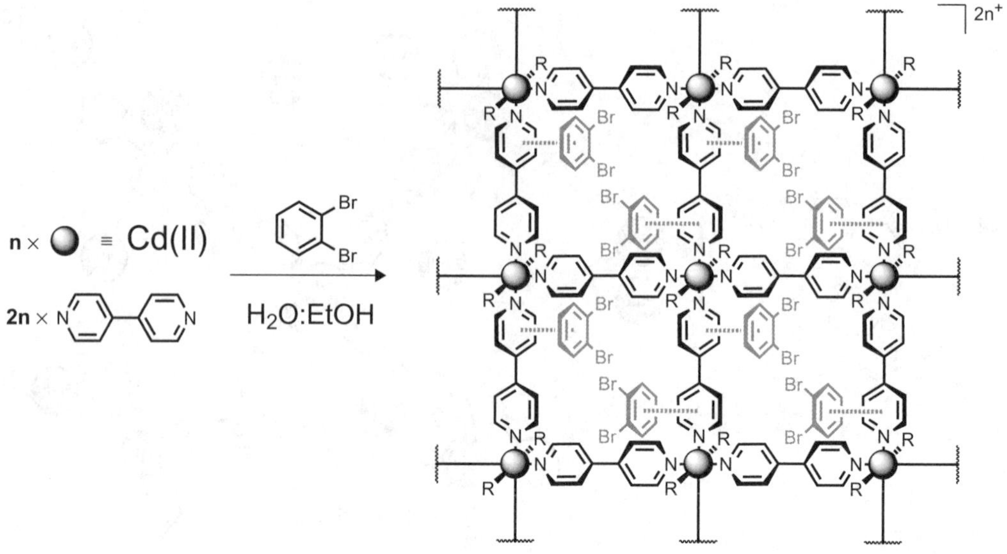

M$_n$L$_{2n}$ coordination polymer grid 4.21

Scheme 4.11 If the metal ions at the corners of a molecular square are not coordinatively saturated then further self-assembly can occur to generate an infinite 2D coordination polymer grid. R = blocking ligand.

blocks enables the properties of a MOF to be easily tuned through structural alterations. Sometimes one may see the term 'coordination polymer' used interchangeably with MOF; both relate to an extended structure based on metal ions connected through organic ligands to form an infinite chain but MOFs themselves fulfil additional criteria of being three-dimensional, crystalline, robust, and (potentially) porous.

MOF synthesis typically involves heating the metal and ligand starting materials in polar solvents such as dimethylformamide in an oven to high temperatures to form the most thermodynamically stable species. Their industrial applications are now driving research into new large-scale, high-yielding preparations under greener, more economic conditions such as aqueous media or even the absence of solvent altogether. After reaction, centrifugation and crystallization are used to isolate MOF crystals of sufficient quality for single crystal X-ray diffraction structural analysis. As in all metal-directed self-assembly processes, the formation of an amorphous material is generally avoided through use of labile ligands and high temperatures that ensure rearrangement to form a single thermodynamically favoured product. Alternatively, if kinetically inert metals or highly charged species are used crystallinity can be improved through the use of additives known as spectator ligands. MOF stability is enhanced by matching metal and ligand sets using the classical principle of hard/soft–acid/base theory with coordinate bonding often augmented by π–π aromatic stacking interactions between ligands.

An example of a MOF is **4.22**, in Figure 4.10. This framework illustrates a common feature of MOFs; the repeating structural motif is not a simple mononuclear metal coordination complex but a larger cluster arising from the assembly of a number of metal ions and ligands. This cluster still has a well-defined geometry, in this case Zn(II) acetate octahedra that occupy the vertices of

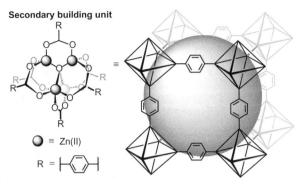

Secondary building unit

◯ ≡ Zn(II)

R = ▭─◯─▭

Metal–organic framework (MOF) 4.22

Figure 4.10 This MOF is a cubic array comprised of octahedral secondary building units connected by benzene dicarboxylate ligands. The sphere indicates the large volume of free space inside a resulting pore that can be exploited for guest uptake.

a cubic lattice, bridged by benzene dicarboxylate ligands to generate the overall structure of the MOF. These multi-metal clusters are often referred to as secondary building units and provide a range of structural geometries such as cubic or tetrahedral.

Secondary building units in MOFs are larger than the mononuclear complexes prevalent in supramolecular cages, which generates cavities of empty space that can be many nanometres in size. MOFs are therefore unique amongst solid-state materials in having some of the lowest known densities, sometimes even lower than liquid water, all because of this extremely high porosity (up to 90% free volume). However, if the spacing between metal clusters becomes too large then a secondary MOF can form within the original assembly to occupy this void space. This is referred to as interpenetration and can be circumvented by filling the cavities with large, strongly bound guest molecules such as counterions during the self-assembly process or by using bulky ligand sets.

The use of chiral building blocks as organic ligands has led to the development of chiral MOFs. For example the secondary building unit shown in Scheme 4.12 uses a chiral ligand based on a D-tartaric acid derivative to construct a trigonal prismatic Zn(II) complex **4.23**. Self-assembly generates a framework that consists of two-dimensional layers of edge-sharing hexagons with the trinuclear metal complex at each corner (**4.24**). At the centre of each hexagon is a chiral pore that is able to preferentially bind one enantiomer of the metal complex $[Ru(bipy)_3]^{2+}$ over another. As such, the stacking of layers on top of one another generates chiral channels that can be used to separate the enantiomers of these metal complexes when a racemic mixture is passed through the material.

Main group self-assembly

In Section 4.2, we saw an example of an octahedral Ga(III) metal complex used for self-assembly. However, main group cations can possess more unusual coordination geometry preferences such as see-saw (disphenoidal), square pyramidal, and trigonal bipyramidal. This provides an opportunity to use them for the self-assembly of structures with contrasting shapes and properties to transition metal assemblies. For example, from Group 16, Te(IV) ions can form a trinuclear Te_3L_3 macrocycle (**4.25**) that resembles an equilateral triangle upon self-assembly with 1,2-benzenedicarboxylate

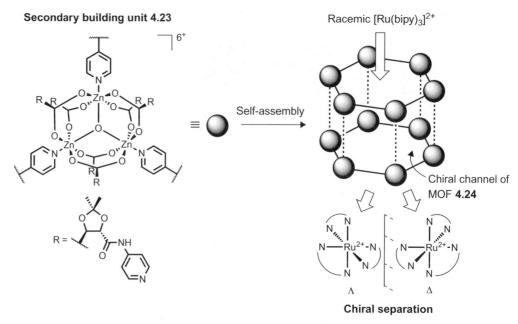

Secondary building unit 4.23

Racemic [Ru(bipy)$_3$]$^{2+}$

Self-assembly

Chiral channel of MOF **4.24**

Chiral separation

Scheme 4.12 The self-assembly of chiral trinuclear Zn(II) secondary building units forms a framework material with chiral channels for enantiomeric separation of [Ru(bipy)$_3$]$^{2+}$ complexes.

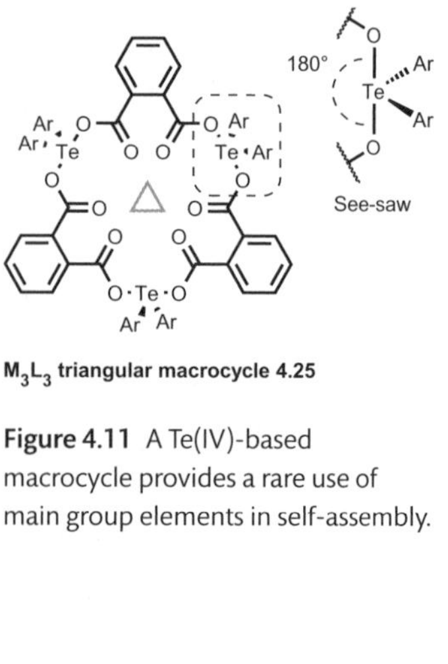

M$_3$L$_3$ triangular macrocycle 4.25

Figure 4.11 A Te(IV)-based macrocycle provides a rare use of main group elements in self-assembly.

ligands (Figure 4.11). The coordination geometry around the Te(IV) ions is the see-saw shape which ensures the O–Te–O bond angle is near 180° for the triangle's edges while the ligands deliver a 60° bite angle for the corners. The formation of a robust cyclic species was confirmed by a single resonance in the ^{125}Te NMR spectrum that was persistent at temperatures of up to 80 °C.

Alkali metal self-assembly

In contrast to d- and p-block cations, the use of s-block metal ions for self-assembly is significantly more challenging due to the difficulties of controlling ionic interactions that are intrinsically non-directional and act over longer distances (i.e. multiple co-ordination spheres). However, metal ions such as Na$^+$ and K$^+$ are prevalent in Nature and provide worthy templates for the construction of biocompatible superstructures by the self-assembly of organic ligand building blocks. This has been demonstrated using Na$^+$ to drive the self-assembly of macrocycles into tubes and larger columnar structures that mimic those of naturally occurring fibres such as collagen (Scheme 4.13). This process is also representative of the way natural alkali metal ion channels are assembled from building blocks of guanine-rich DNA segments. The self-assembly of macrocycle **4.26** occurs because bound Na$^+$ interacts with more than one ring at a time. This enables the formation of oligomers and then polymers as observed by mass spectrometry (Scheme 4.13). However, the introduction of bulky triptycene substituents to the catechol groups in the macrocycle (**4.27**) blocks self-assembly on steric grounds. Over time, the polymers of macrocycle **4.26** assemble into tubes and then larger fibrous structures due to long-range electrostatic interactions between components. Therefore, this system is replicating biological self-assembly because there is a continual increase in the extent of structural organization over several length scales.

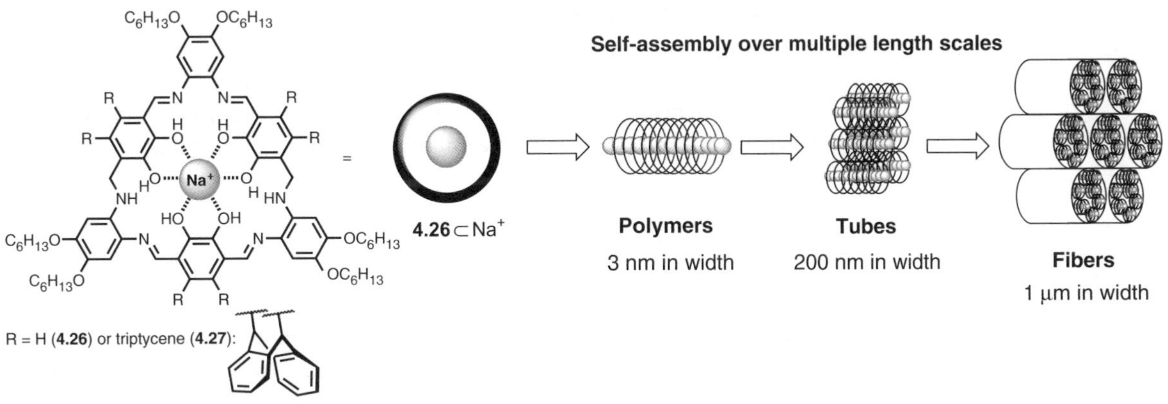

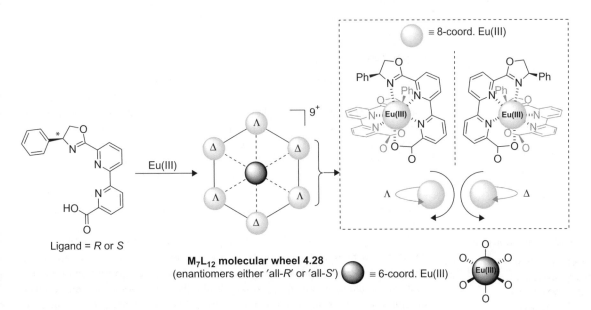

C₆H₁₃O   OC₆H₁₃

R = H (**4.26**) or triptycene (**4.27**):

4.26 ⊂ Na⁺

Self-assembly over multiple length scales

Polymers
3 nm in width

Tubes
200 nm in width

Fibers
1 μm in width

Scheme 4.13 A crown ether-like macrocycle forms molecular architectures of increasing size by self-assembly with Na⁺ cations.

Lanthanide metal self-assembly

Like s-block metals, the variable and high coordination numbers and geometries of the lanthanide metal ions make their use in the design of self-assembled supramolecular architectures challenging. However, there are examples of lanthanide ion templated macrocycles, helicates, cages, and even MOFs analogous to those constructed using transition metal ions. The functionality of these assemblies benefit from the unique luminescent and magnetic properties of the f-block ions they contain. An interesting structure prepared via the self-assembly of seven Eu(III) ions and twelve bi-pyridyl containing ligands functionalized with additional carboxylate and oxazoline donor groups to satisfy the large coordination number of the metal cation is shown in Scheme 4.14. Characterized by X-ray crystallography, this unusual M_7L_{12} assembly (**4.28**) resembles

Ligand = *R* or *S*

M₇L₁₂ molecular wheel 4.28
(enantiomers either 'all-*R*' or 'all-*S*')

≡ 6-coord. Eu(III)

≡ 8-coord. Eu(III)

Scheme 4.14 A molecular wheel comprised of six peripheral eight-coordinate Eu(III) complexes and one six-coordinate Eu(III) complex in the centre. Stereoisomerism arises through point chirality (ligand) and axial chirality (lanthanide metal complex).

a molecular wheel in which the axial chirality (Λ or Δ) of the six peripheral Eu(III) complexes alternates as one moves around the ring. Furthermore, since the ligand itself also exhibits point chirality (*R* or *S*), the heptameric assembly **4.28** can be formed as two enantiomers i.e. the twelve ligands are either 'all-*R*' or 'all-*S*'. As a result, the f-metal luminescence from an enantiopure assembly is circularly polarized which provides a method for the sensing of chiral substrates.

4.3 Neutral self-assembly

It is clear from the previous sections discussed above that metal cation–ligand co-ordinate bonding is a powerful template for constructing complex supramolecular structures by self-assembly. However, alternative neutral organic species can also be exploited to direct the assembly process.

This section begins by demonstrating how arrays of hydrogen bonds provide an effective tool for constructing some impressive synthetic systems, in analogous fashion to the self-assembly of DNA. The stability of DNA is augmented through π–π stacking (Figure 4.1), and herein we also highlight examples where, by careful consideration of the solvent medium, these interactions can be exploited for the self-assembly of aromatic molecules. Finally, to realize true biomimetics, self-assembly must occur under biological conditions, namely salty water at moderate temperatures (~37 °C). This inevitably requires an appreciation for the influence of the hydrophobic effect which is reintroduced in the latter examples. Doing so enables assemblies to be formed in water across length scales from molecular capsules to macromolecular micelles to framework materials.

While there are many parallels between DNA and the examples below it should be noted that individual strands of deoxyribonucleic acid have themselves been exploited as building blocks for the self-assembly of rationally designed nanostructures. Further details of this now widespread technique, known as DNA origami, are beyond the scope of this book but links to resources are provided in Section 4.7.

Hydrogen bonding

In the absence of strong electrostatic intermolecular forces, the self-assembly of neutral molecules is commonly driven by hydrogen bonding (HB) because of its reasonable strength, directionality, orientation, and specificity (see Chapter 1, Section 1.3). The variety of functional groups capable of acting as hydrogen bond donors or acceptors enables numerous hydrogen bonds to act in concert. The importance of this is demonstrated by the fact that the equilibrium constant for the self-association of a simple amido pyridine derivative **4.29** is only 13 M^{-1} in CHCl$_3$, an aprotic solvent (Scheme 4.15). Notably, in a protic solvent such as methanol the dimer **4.30** is not formed due to competition with the solvent's strong hydrogen bonding capability.

As shown in Scheme 4.15, the HB donor (amide) and acceptor (pyridine) groups can be ascribed D and A labels respectively (see also Chapter 3, Section 3.2). This can be very useful when designing and analysing the patterns of larger hydrogen bond networks containing multiple donor and acceptor units for the self-assembly of more complex structures. Trimer systems composed of imide (ADA) functionalities are common although, as the system becomes more complex, repulsive secondary interactions between spectator oxygen atoms can lead to an energetic penalty (Figure 4.12).

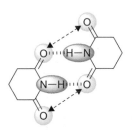

Imide–imide interactions

Figure 4.12 Repulsive secondary interactions (dotted arrow) can occur if not all the hydrogen bonding requirements are satisfied.

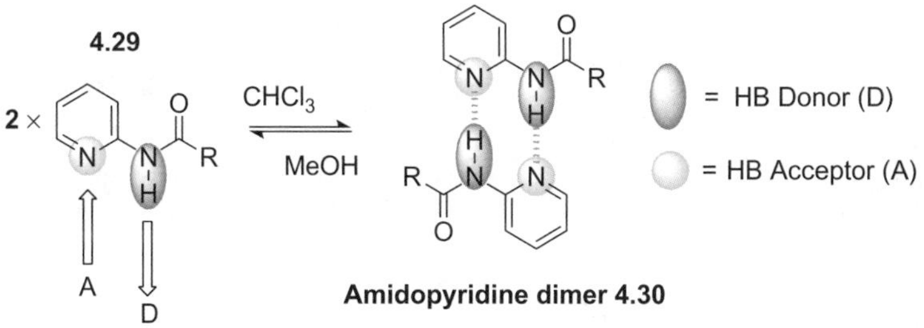

Amidopyridine dimer 4.30

Scheme 4.15 The solvent dependency of a hydrogen bond duplex formed between two amidopyridine molecules.

Cyanuric acid and melamine molecules are excellent building blocks for hydrogen bond self-assembly. A 1:1 stoichiometric mixture forms extended supramolecular assemblies in which the HB requirements for each sub-unit are completely satisfied (Scheme 4.16). The formation of an infinite polymeric assembly **4.31** causes precipitation from solution. To prevent this, the HB network was restricted by replacing the peripheral hydrogen bonding groups with solubilizing alkyl and aryl groups, forming a smaller assembly **4.32** called a rosette (Scheme 4.16).

Hydrogen bond self-assembly is not restricted to the solution phase but can also be used to create ordered molecular patterns on surfaces, as highlighted by the arrangements of perylene diimide and melamine building blocks on gold in Figure 4.13. In **4.33** the two-dimensional hydrogen bond network generates a honeycomb structure in

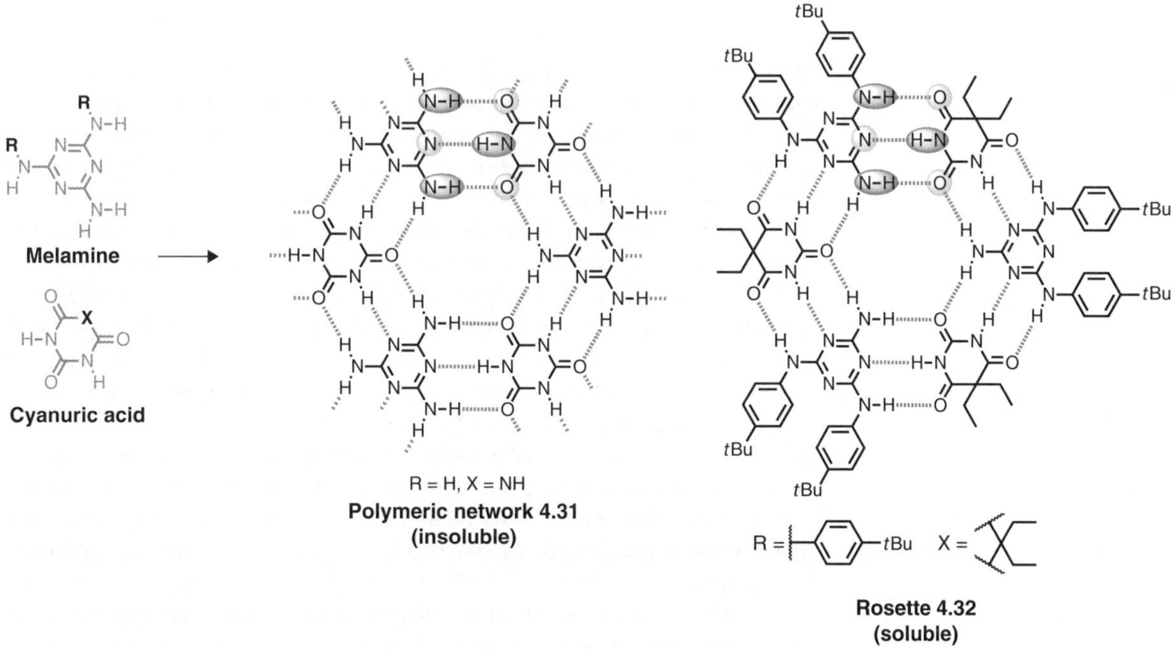

Scheme 4.16 Formation of molecular rosettes by hydrogen bond-driven self-assembly.

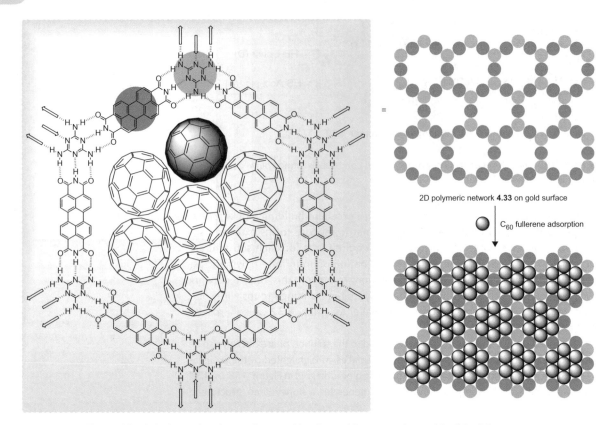

2D polymeric network **4.33** on gold surface

C_{60} fullerene adsorption

Figure 4.13 Self-assembly of a hydrogen bond network on a gold surface guides a second assembly of C_{60} fullerenes.

which the nanometre-sized pores are large enough to accommodate clusters of up to seven C_{60} fullerene molecules. In essence, one self-assembled network acts to template the adsorption of another. Further control can be gained by removing the melamine spacer unit which shrinks the pore size and so encapsulation of only a single fullerene in each of the hydrogen bond-framed windows is possible. The ability to assemble and tune these frameworks on surfaces has important implications for the construction of molecular memory storage devices that require precise organization on the nano scale.

Back in solution, hydrogen bond self-assembly can afford discrete macrocyclic species. For example, complementary hydrogen bonding between carboxylic acids in DMSO has been employed to direct the assembly of the four-component molecular square **4.34** in quantitative yield (Scheme 4.17). Under thermodynamic control, the final binding event is intramolecular and is entropically favoured over any intermolecular binding that would generate longer oligomeric chains. However, this product ratio can be reversed if the shape of the bis-carboxylic acid is altered. Photoisomerization of the *cis*-azobenzene spacer unit to a *trans* configuration prevents any ring-closing reaction and so one-dimensional linear chains are assembled instead (**4.35**).

Beyond macrocycles, hydrogen bonding self-assembly has been extended to the formation of three-dimensional cages. These structures act like capsules in which

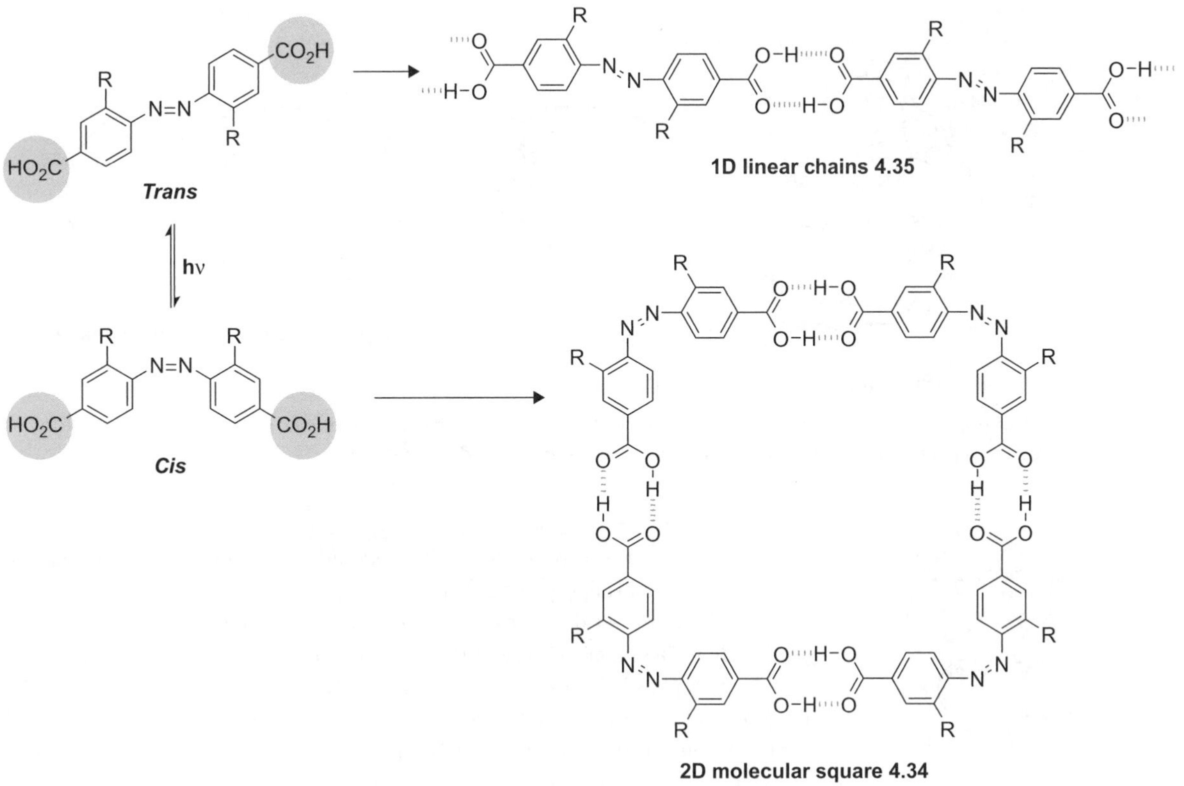

Scheme 4.17 The geometry of the bis-carboxylic acid building block dictates the product of self-assembly.

the guests inside their cavities are completely shielded from surrounding solvent molecules allowing them to mimic biological assemblies such as enzymes or membrane channels. For biocompatible applications, the synthetic cages must be able to self-assemble in water. As mentioned earlier, this presents a significant challenge if only relatively weak hydrogen bonding between neutral species are to be employed. This makes the self-assembly of dimeric cavitand capsule **4.36** under aqueous conditions particularly impressive (Scheme 4.18). Critical to its stability are the 16 hydrogen bonds shared between neutral benzimidazolone groups that form a cyclic seam to hold the two cavitand hemispheres together. The cage is able to encapsulate organic guests such as stilbenes that would not normally dissolve in water suggesting a potential application as a vector for hydrophobic drug molecules. More recently, peptides have been appended to macrocyclic cavitand scaffolds enabling their self-assembly into dimers by hydrogen bonding arranged in β-barrel type binding motifs between the peptidic backbones.

Self-recognition and complementary hydrogen bonding have also been used to assemble a three-dimensional structure that resembles a tennis ball (Scheme 4.19). Again, the hollow interior of **4.37** can be used to capture guest molecules such as methane through van der Waals interactions. Interestingly, ^{1}H NMR spectroscopy reveals two sets of signals for free and encapsulated guest when the gas is bubbled through a solution of the host. This indicates that the rate at which methane molecules

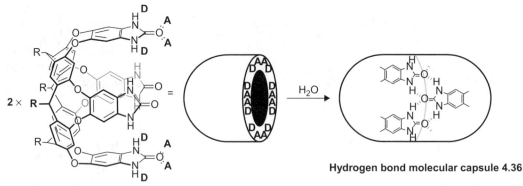

Hydrogen bond molecular capsule 4.36

Benzimidazolone-based cavitand

Scheme 4.18 Two bucket-shaped cavitand macrocycles decorated with benzimidazolone groups self-assemble into a capsule in pure water. D = hydrogen bond donor, A = hydrogen bond acceptor.

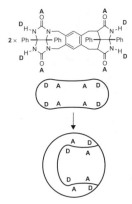

Hydrogen bond tennis ball 4.37

Scheme 4.19 Formation of a molecular tennis ball by complementary hydrogen bonding.

inside exchange with those outside is relatively slow (< 300 times per second at room temperature), which is a consequence of the strong seam of hydrogen bonds having to (partially) unzip to allow guest exchange.

It should be noted that capsule self-assembly is not just restricted to hydrogen bonding. Taking advantage of the strict requirements for bond linearity, halogen bonding has been exploited to assemble two cavitand hemispheres (**4.38**). In agreement with the trend in halogen bonding described in Chapter 2, the most stable capsules result when the halogen bond donor contains an iodine or bromine atom, with no cages formed by the chlorine- or fluorine-containing systems (Figure 4.14).

The previous examples all made use of *inter*molecular hydrogen (or halogen) bonding. However, the occurrence of these non-covalent interactions between donor/acceptor motifs within the same molecule, i.e. in an *intra*molecular fashion, can influence the structure's conformation. This concept will be familiar from how the secondary structures of proteins (e.g. α-helices or β-sheets) are formed by patterns of intramolecular hydrogen bonds between amino acid residues. Abiotic molecules that mimic this ability are called **foldamers**. For example, hydrogen bonding between neutral benzene–acylurea monomers in an oligomeric amide chain triggers folding into an α-helix conformation (Scheme 4.20). Measuring up to 4 nm in length with seven helical turns, foldamer **4.39** is, in fact, longer than the average protein α-helix, (ten residues with three turns). By contrast, the phenyl amide sub-units of the material Kevlar only allow *inter*molecular hydrogen bonding between polymer chains, instead giving planar sheet-like structures.

Aromatic stacking interactions

Aromatic stacking interactions are often weaker than hydrogen bonds making their use in self-assembly more challenging. As we will see in the following examples, the choice of solvent is critical in determining the strength and nature (i.e. potential for electrostatic donor–acceptor character) of aromatic stacking interactions. This is particularly important when these π–π interactions are the sole driving force for self-assembly and are not, as we have seen previously, supplementing other non-covalent interactions.

The solvent dependence of aromatic stacking interactions has been investigated via the self-assembly of oligomeric chains consisting of alternating dialkoxynaphthalene (DAN) and naphthalene diimide (NDI) units (called aedamers, **4.40**). In solution, these aedamers assemble into folded structures driven by intramolecular aromatic stacking between the DAN and NDI motifs (Figure 4.15). While both these species have no formal charge, the electron donating or withdrawing groups of DAN and NDI respectively mean that, relative to one another, they are considered 'electron rich' or 'electron poor'. As such, interactions between these π surfaces of electronic complementarity, known as donor–acceptor aromatic stacking, can provide a significant driving force for forming the face-to-face assembly. However, it was found experimentally that increasing the polarity of the solvent, which one might expect to weaken the donor–acceptor electrostatic interactions, instead increases the free energy of the DAN–NDI association leading to a more robust folded assembly. Therefore, this trend reflects a decreasing degree of solvation of the lipophilic aromatic surfaces in solvents of increasing polarity. This 'solvophobic' contribution is the other major driving force for aromatic stacking interactions and, with clear similarities to the hydrophobic effect (see later in this chapter), tends to dominate under polar solvent conditions.

So when does donor–acceptor electronic complementarity become important for aromatic stacking interactions? Investigations into the solvent dependence of self-assembly of another aromatic molecule, perylene diimide (PDI, **4.41**), has shown the answer to be aprotic solvents of low polarity, such as alkanes and ethers. Under these conditions, a reversal in the previous solvent polarity trend is observed and so the aromatic stacking can be considered to be more electrostatic in nature (Figure 4.15). PDI has a large, planar, and quadrupolar aromatic surface allowing it to form impressive twisted columnar aggregates via intermolecular π–π stacking in solution. The strong colour of these dye molecules allows their self-assembly to be easily monitored by UV-vis spectroscopy or even by the naked eye.

An experimental methodology has been devised to quantify other types of weak non-covalent interactions involving aromatic rings and to elucidate their influence on self-assembly. Specifically, the strength of edge-to-face aromatic interactions between phenyl rings, otherwise known as CH–π interactions, were determined by

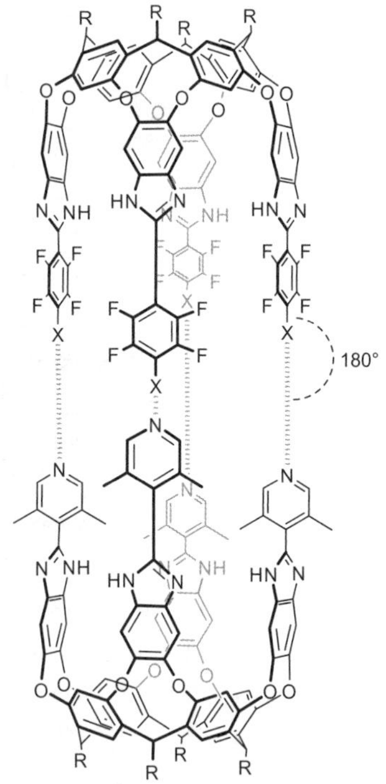

Halogen bond molecular capsule 4.38
X = Br, I

Figure 4.14 Halogen bonding is used in the assembly of dimeric cavitand capsules.

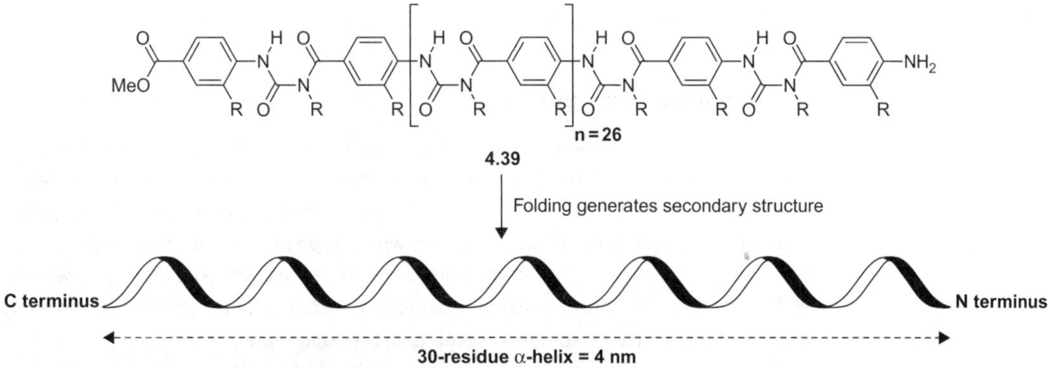

Scheme 4.20 Intramolecular hydrogen bonding within an aryl-acylurea oligomer generates a folded secondary structure that resembles a protein α-helix.

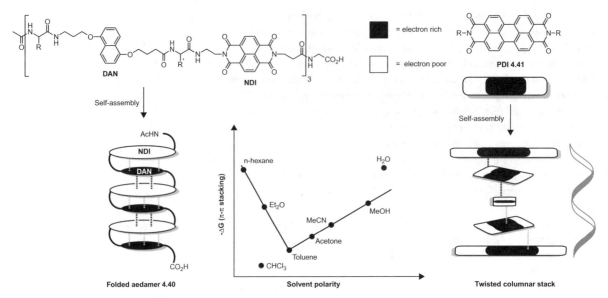

Figure 4.15 Experimental studies on the solvent dependency of aromatic stacking interactions using self-assembly of an aedamer (left) and perylene diimide (right). Electrostatic and solvophobic contributions vary within the two solvent regimes. Plot adapted with permission from M. S. Cubberley *et. al.*, *J. Am. Chem. Soc.*, 2001, **123**, 7560. Copyright (2001), American Chemical Society.

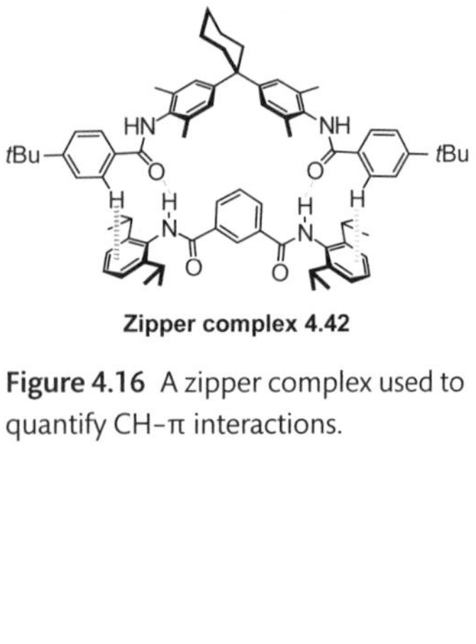

Zipper complex 4.42

Figure 4.16 A zipper complex used to quantify CH–π interactions.

using a range of molecular 'zipper' complexes. While the full details of this method can be found in the relevant article listed in Section 4.7, the basic premise was to sequentially alter structural components involved in the non-covalent interaction of interest and examine the effect on the free energy of zipper complex formation. From the system based on complex **4.42** (Figure 4.16) the free energy of one terminal edge-to-face interaction was determined to be –1.3 kJ mol^{-1} (i.e. a favourable interaction). Albeit weaker than for example hydrogen bonds (> 10 kJ mol^{-1}), these interactions are known to be important in the self-assembly of polycyclic aromatic hydrocarbons which, as we will see later, can have a strong influence on their properties as electronic materials.

Supramolecular polymers

We have seen how supramolecular self-assembly can be used to dictate the conformation of covalent polymers (e.g. α-helix foldamers, Scheme 4.20) however, supramolecular chemistry can also be used to non-covalently link monomers together producing a unique class of assemblies known as '***supramolecular polymers***'. In a supramolecular polymer, arrays of monomeric units are brought together by reversible and highly directional non-covalent interactions, resulting in polymeric properties in dilute and concentrated solution as well as in the solid-state.

Hydrogen and halogen bonding, the hydrophobic effect and π–π aromatic stacking have all been exploited as non-covalent interactions in the self-assembly of supramolecular polymers. An example of the latter uses the strong binding of the electron poor aromatic trinitrofluorenone by a receptor composed of two electron rich

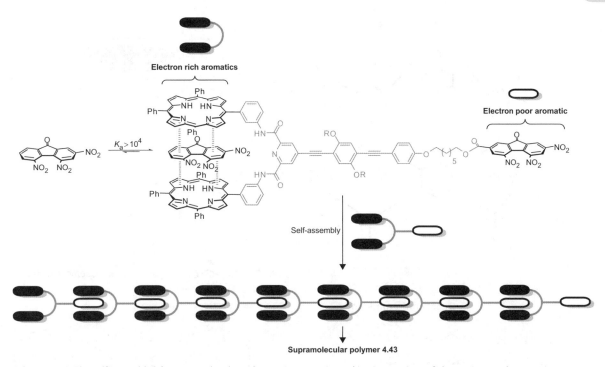

Scheme 4.21 The self-assembly of a supramolecular polymer via aromatic stacking interactions of electronic complementarity.

porphyrin units (Scheme 4.21). In a relatively low polarity solvent, there is a strong donor–acceptor contribution towards the aromatic stacking interactions and a large association constant for the 1:1 host–guest complex (K_a = 42 000 M^{-1}). Integrating both these moieties into the same molecular scaffold produces a heteroditopic monomer that self-assembles into supramolecular polymer **4.43** through head to tail association (i.e. self-recognition).

Self-assembly in water: the hydrophobic effect

Robust and controlled self-assembly (and even disassembly) under aqueous conditions is critical when seeking to apply synthetic supramolecular systems within the fields of biology and medicine. An earlier example (Scheme 4.18) demonstrated that hydrogen bonding can be used to drive self-assembly in water; however, this is relatively rare and instead the majority of systems primarily exploit the hydrophobic effect. The origins of this phenomenon for host–guest recognition have been explained in Chapters 2 and 3.

The following examples will highlight the potency of the hydrophobic effect for the self-assembly of neutral organic species which, due to their low polarity, are inherent hydrophobes. We will see how, under aqueous conditions, the hydrophobic effect facilitates the formation of a range of different structures across various length scales, from molecular complexes to macromolecular constructs such as micelles, nanotubes, and materials such as gels.

First introduced in Chapter 3, water-soluble deep cavity cavitands (**4.44**) have been employed for investigating hydrophobic-driven molecular self-assembly in

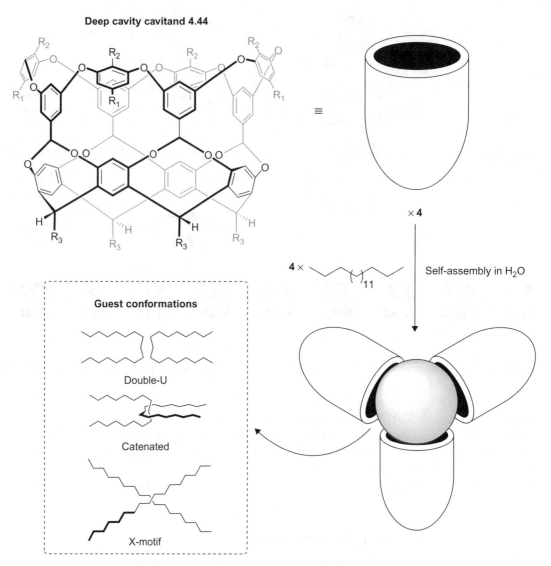

Deep cavity cavitand 4.44

Self-assembly in H$_2$O

Guest conformations

Double-U

Catenated

X-motif

Scheme 4.22 The self-assembly of a tetrahedron of four deep cavity cavitands in water is driven by hydrophobic encapsulation of C$_{17}$ n-alkyl chains.

water. These hosts are known to bind a range of acyclic aliphatic guests, enabling their self-assembly into well-defined supramolecular entities via the hydrophobic effect (Scheme 4.22). For example, long chain alkanes have been shown to trigger the formation of tetrameric complexes with a 4:2 host–guest stoichiometric ratio in order to satisfy the complete encapsulation of these hydrophobes in water. As indicated by the symmetric nature of the ^{1}H NMR spectrum, these complexes have a pseudo-tetrahedral geometry meaning that the flexible n-alkyl chains must be assembled into one of the three different arrangements seen in Scheme 4.22. This demonstrates that, by deploying the hydrophobic effect, non-preorganized guests with almost no functional handles can act as effective templates to direct self-assembly under aqueous conditions. More extensive details on the thermodynamic driving forces behind hydrophobic self-assembly are available from the resources listed in Section 4.7.

Micelles are large supramolecular aggregates composed of a single molecular component called a surfactant. For self-assembly in water, a surfactant molecule must have well-defined hydrophobic and hydrophilic regions to its structure, i.e. an amphiphile. Depending on the conditions and concentration of the aqueous solution, surfactant molecules may also assemble into other ordered structures such as monolayers, bilayers, and vesicles (e.g. the virus in Scheme 4.2).

Fullerenes have been exploited as building blocks for micelles because they provide a lipophilic spherical scaffold from which multiple hydrophilic (carboxylic acids) and lipophilic groups (long alkyl chains) may be appended (Scheme 4.23). Upon dissolution in water at pH 7, the hydrophobic effect causes thousands of these amphiphiles to self-assemble into large rod-shaped micelles **4.45** that can be characterized by electron micrography. This imaging technique was also used to show that, under basic conditions, deprotonation of carboxylic acid groups introduces negative charges that drastically alter the micelle structure, an effect which could be exploited for drug-delivery systems.

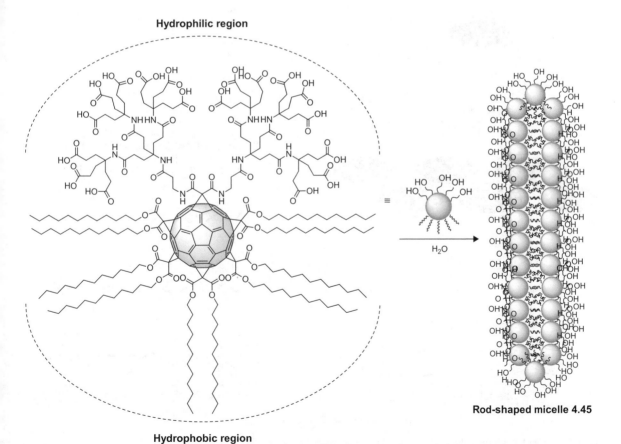

Scheme 4.23 C_{60} fullerene amphiphiles self-assemble into rod-shaped micelles in water.

The library of naturally occurring amino acids provide a range of charge, hydrophobicity, and size that influence the physiochemical properties of the resulting protein. In Nature, this will also dictate the self-assembly of these biological macromolecules into even larger multicomponent constructs such as antibodies, ribosomes, and the heme protein assembly.

Taking inspiration from this, chemists have covalently linked specific amino acids to give new synthetic polypeptides designed to self-assemble into large, functional constructs and generate materials that are biocompatible and biodegradable. For example, synthetic macrocycles composed of alternating D- and L-amino acids self-assemble into polymeric nanotubes (**4.46**) through the stacking of neutral rings on top of one another by intermolecular hydrogen bonding (Scheme 4.24). While polypeptides will typically rely on intramolecular hydrogen bonding to define their secondary

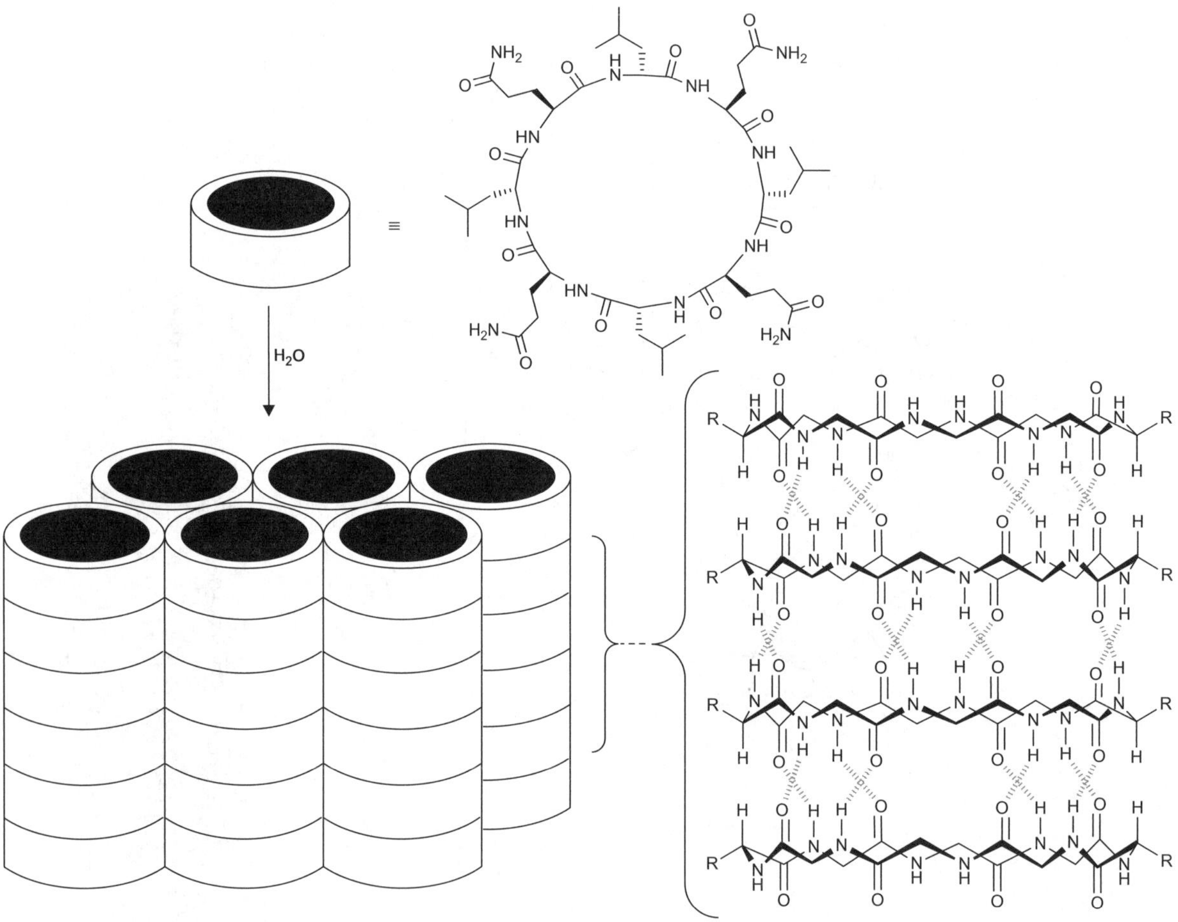

Peptide macrocycle nanotubes 4.46

Scheme 4.24 Nanotubes formed through self-assembly of peptide macrocycles.

structure, here intermolecular supramolecular interactions dominate due to the unique cyclic structure. The tubes are hollow and their diameter can be adjusted via the number of constituent amino acids in the macrocycle monomer. One potential application is their use for antibacterial activity by assembling them in bacterial membranes to increase permeability.

Supramolecular gels typically use low molecular weight molecules (gelators) that, in solution, self-assemble into long supramolecular polymer fibre-like chains and then associate strongly with one another via non-covalent cross linking. This generates a robust solid network propagated throughout the liquid component, which must remain mobile to exhibit the properties of a gel. The classical test for a gel is to invert its vessel and observe if the solvent is able to support its own weight without collapse.

Dibenzylidene-D-sorbitol (DBS) was one of the first known low molecular weight supramolecular gelators (Scheme 4.25). Self-assembly of these neutral, amphiphilic molecules is caused by intermolecular aromatic stacking interactions between the phenyl rings. Gelation was for a long time restricted to organic solvents on account of poor aqueous solubility (**4.47**). However, the appendage of hydrazide groups to each of the phenyl rings of DBS (**4.48**) made it capable of forming a gel in water (known as a hydrogel). The success of this strategy relies on achieving the correct balance between water solubility and hydrophobicity, to ensure self-assembly occurs. Importantly, **4.48** is able to form hydrogels across a large pH range because the hydrazide group remains neutral throughout. This pH-independent structure may prove key to the application of these materials in tissue engineering, separation science, and sensing.

Crystal engineering

The strategic use of specific, directional non-covalent interactions in the solid-state can create anisotropic materials with novel electronic, magnetic, optical, and physical properties. The crystallization of a compound from solution is essentially a form

Dibenzylidene-D-sorbitol gelator (DBS)

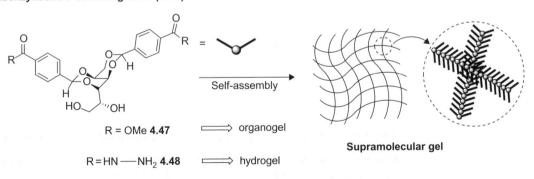

R = OMe **4.47** ⟹ organogel

R = HN—NH$_2$ **4.48** ⟹ hydrogel

Scheme 4.25 Self-assembly of supramolecular gelator DBS to form gels in either water or organic solvents. DBS has a sorbitol 'backbone' and bent phenyl 'wings' and so the conformation is said to resemble a butterfly.

of solid-state self-assembly in the sense that, during the process, molecules interact with one another and gradually pack to form a new structure of optimum orientation. The need to control this process has led to the development of a highly active field of research known as crystal engineering. By identifying crucial links between a molecule's structure and properties at a range of length scales, new materials can be rationally designed and fabricated with tailor-made functionalities. Without appreciating the influence of these intermolecular interactions, controlling self-assembly in the solid-state relies on trial, error, and serendipity. Alongside experimental work, computational chemistry is now providing great assistance in predicting the outcome of self-assembly processes.

A pertinent example of the structure–property relationship is the influence of molecular packing on charge transport in semiconducting crystals composed of small organic molecules. In general, planar aromatic molecules such as tetrathiafulvalene (TTF) derivatives will adopt either a bricklayer (face-to-face) or herringbone (edge-to-face) packing arrangement, with the former enhancing intermolecular π orbital overlap required for conductivity (Scheme 4.26). As such, the difference in charge mobility between TTFs **4.49** and **4.50** is explained by their contrasting crystal structures; reducing the number of potential intermolecular S–S contacts (dispersive forces) in the pentyl-based TTF (**4.49**) enables π–π stacking to dictate self-assembly instead.

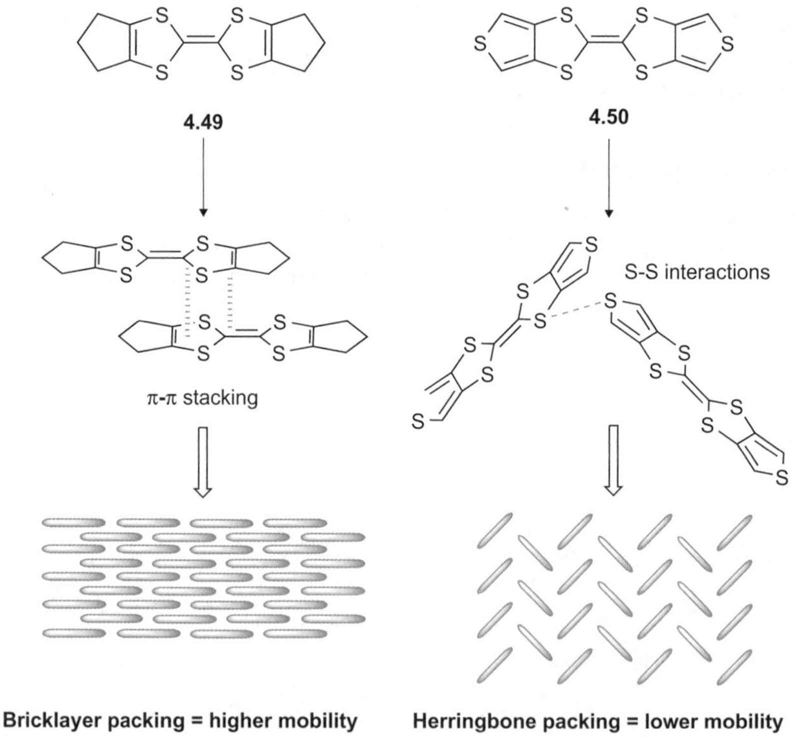

4.49   **4.50**

π-π stacking   S-S interactions

Bricklayer packing = higher mobility   **Herringbone packing = lower mobility**

Scheme 4.26 Intermolecular interactions are key to understanding the connection between molecular structure, crystal packing, and properties of TTF electronic materials.

Precise crystal engineering presents a significant challenge if the only inter-molecular interactions available for exploitation are extremely weak, e.g. van der Waals forces. However, the use of strong metal–ligand coordinate bonding in metal–organic frameworks (MOFs) introduced in Section 4.2 provides a clear demonstration that self-assembly can be performed in the solid-state. MOFs can be considered part of a broader class of self-assembled materials known as supra-molecular organic frameworks. Another group in this family are hydrogen-bonded organic frameworks (HOFs), porous materials assembled through hydrogen bonding. These bonds assist in controlling crystallization, however, with neutral species the framework is prone to collapse on removal of guest/solvent molecules from the pores if the hydrogen bond network is not robust enough.

A recognized approach to stable HOFs exhibiting permanent porosity is to make use of motifs known to form strong hydrogen bond networks and synthetically mod-ify the constituent molecules so that crystallization generates pores. For example, the 4,5-disubstituted benzimidazolone group we saw in Scheme 4.18 forms a planar rib-bon-like network via hydrogen bond directed self-assembly. Covalently appending these groups to the three spokes of a rigid triptycene scaffold ensures that crystals with one-dimensional channels are formed via intermolecular hydrogen bonding, the strength of which ensures pores can be evacuated (Scheme 4.27). This endows mate-rial **4.51** with a large surface area, allowing it to adsorb gases such as CO_2 and CH_4. Interestingly, uptake of the former is favoured because of a stronger attraction with the polar benzimidazolone surface.

Whilst MOFs, and even covalent organic frameworks, are considered superior materials for gas storage on the criterion of robustness, the way in which HOFs are made gives them an edge in terms of sustainability. Crystallizations are typically performed at lower temperatures and in more environmentally friendly solvents. Furthermore, these materials can be easily recycled by dissolution which is desirable for reusable devices.

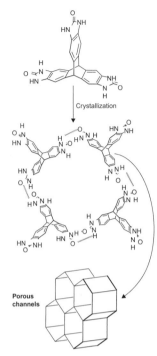

Hydrogen-bonded organic framework 4.51

Scheme 4.27 HOFs are porous crystalline materials formed from self-assembled networks of hydrogen bonds.

4.4 Anion self-assembly

In comparison to metal ions, the use of anions in supramolecular self-assembly is relatively underdeveloped, in part because appropriate ligand design has histori-cally lagged behind due to the challenges associated with anion recognition (see Section 2.3). Weak binding and a lack of a well-defined coordination geometry has traditionally restricted anions to playing a supporting role to metals in defining su-pramolecular structure, as highlighted by some examples in Section 4.2. However, the development of a wider range of hosts has improved both anion affinity and selec-tivity for these ligands, while the exploitation of hydrogen and halogen bonding can provide directionality. This has enabled the construction of discrete supramolecular architectures solely with anions used as coordination centres (i.e. primary templates). Anions have even been used as a tool for directing the long-range ordering of mole-cules into networks, generating functional materials such as gels, micelles, or crystal-line arrays for guest capture and release.

Some of the first structures generated through anion self-assembly were fol-damers, in which acyclic host molecules (ligands) wrap themselves around anion guests

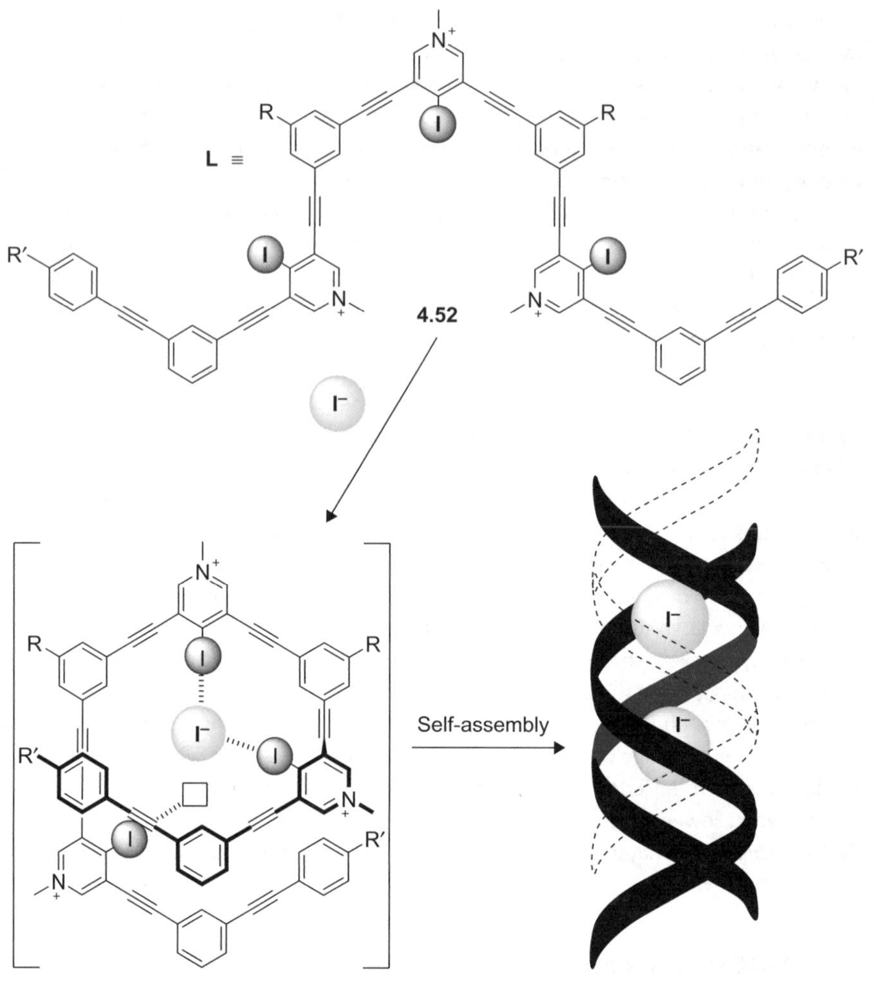

Scheme 4.28 Iodide binding dictates the self-assembly of halogen bond donor ligands into helical over folded structures. The box in **4.53** indicates a vacant coordination site. A = anion, L = ligand.

(coordination centres). For example, iodide recognition by halogen bond host **4.52** can generate the folded secondary structure **4.53** (Scheme 4.28). However, to satisfy the anion coordination requirement of the host, the system was found to further self-assemble into a triple helicate **4.54**, in which two iodide anions are coordinated via eight halogen bonds. The polarized iodine atoms of the electron-deficient iodopyridinium motifs in each ligand strand are potent halogen bond donors, enabling formation of a stable tubular channel through the centre of the helicate. Whilst it should be noted that intermolecular aromatic stacking interactions between the ligands (evident from the crystal structure) provide stability, removal of the anion template causes disassembly and the ligands revert back to the unfolded state.

Similar dinuclear helicate systems have been reported. In one example, a diamino-bis-pyridine ligand is used to generate A_2L_2 helicate **4.55**, upon exposure to HCl$_{(aq)}$ (Scheme 4.29). The acid protonates the pyridyl groups and switches on chloride binding through charge assisted hydrogen bonding interactions.

Self-assembly via reversible covalent bond formation can also be influenced through anion coordination. For example, an aromatic dicarboxylate acts as an anion template for [2+2] macrocycle **4.56** formed using Schiff base chemistry (Scheme 4.30). Since the amide-based anion binding sites are positioned at opposite ends of the macrocycle, the highest macrocyclization yields are found with a *para-* over *meta-*substituted benzene dicarboxylate template to enable geometric complementarity between the cyclic host and anion guest.

Variations on the above have integrated the carboxylate anion template into the aldehyde macrocycle precursor so that the product of the first Schiff base reaction can itself template the cyclization procedure. These self-replicating systems, in which a product acts as a catalyst for its own formation, provide insights into the possible chemical processes that started life on Earth. Further information is provided in the resources in Section 4.7.

The anion PO_4^{3-} is a highly charged tetrahedral species that can accommodate up to twelve hydrogen bonds, making it a potential template for the anion-driven self-assembly of three-dimensional supramolecular cages. The bis-urea ligand shown in Scheme **4.31** forms a 3:1 host–guest complex with PO_4^{3-} with each urea functionality donating

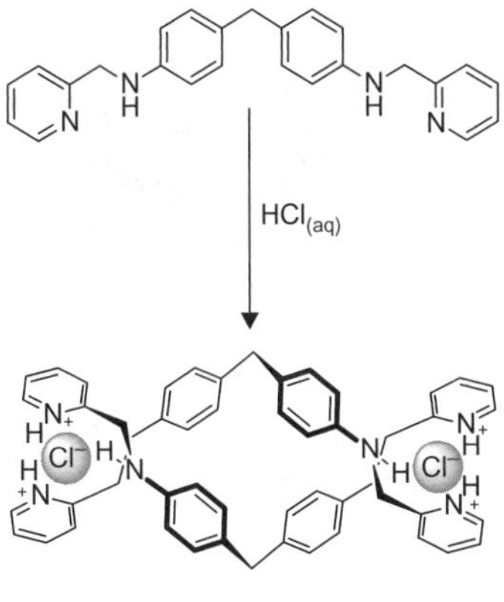

A_2L_2 helicate 4.55

Scheme 4.29 A double helicate formed through chloride recognition.

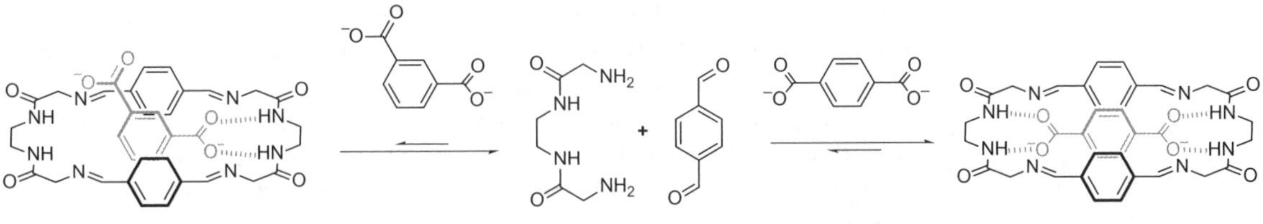

Low yield of macrocycle 4.56

High yield of macrocycle 4.56

Scheme 4.30 Self-assembly of a macrocycle by reversible covalent bond formation is dictated by the binding of an anion guest in methanol.

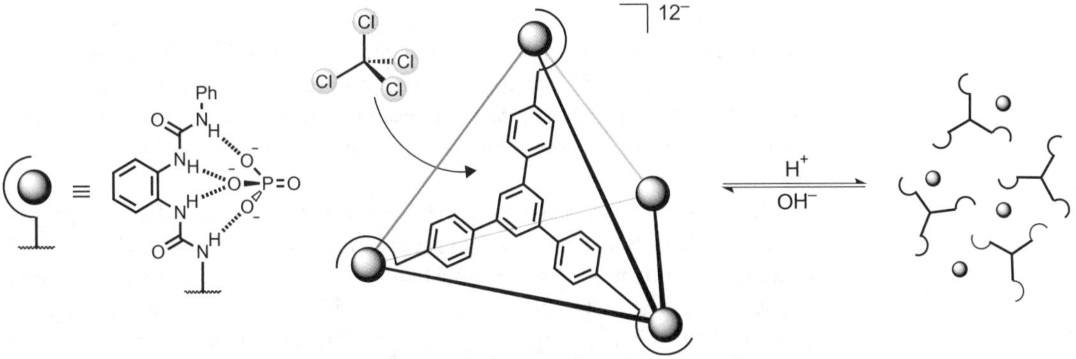

A_4L_4 tetrahedron 4.57

Scheme 4.31 The first example of an anion templated tetrahedral cage.

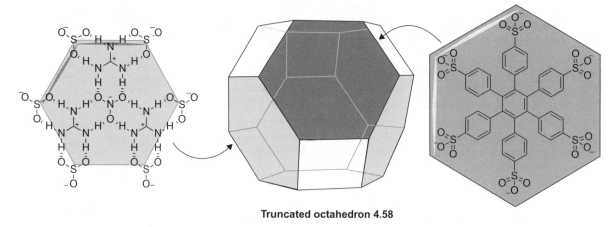

Truncated octahedron 4.58

Figure 4.17 Two sets of complementary tiles self-assemble through hydrogen bond anion coordination to generate an architecture in the shape of a truncated octahedron.

two hydrogen bonds. As in metal-directed assembly, prior knowledge of this binding stoichiometry and geometry was essential in the synthesis of tetrahedral cage **4.57**. A total of 48 hydrogen bonds are formed to generate this A_4L_4 structure in which the PO_4^{3-} anions occupy the vertices of the tetrahedron with urea-based ligands the faces. The central cavity of these anion cages has been exploited for host–guest chemistry; cage encapsulation of complementary shaped tetrahedral guests such as chlorofluorocarbons was monitored by NMR spectroscopy and mass spectrometry. Owing to their pH dependence, reversible anion directed assembly and disassembly of the cage can enable the controlled capture and release of guest species using a switchable pH stimulus.

Akin to the molecular panelling approach introduced in Section 4.2 of this chapter, anion self-assembly has been showcased through the synthesis of a convex polyhedron (**4.58**) in the shape of a 'truncated octahedron' (Figure 4.17). This impressive structure relies on the guanidinium motif as an anion binding basis unit to coordinate tetrahedral sulfonate and trigonal planar nitrate anions which stitch together molecular tiles using a total of 72 hydrogen bonds. This capsule has a huge internal volume to encapsulate large guests. In fact, rare examples of transition metal clusters were found to be stabilized by the confinements of the pore, generating 'ship-in-a-bottle' assemblies.

More recently it has been shown that it is possible to self-assemble an expanded supramolecular organic framework in water using anion templation (Scheme 4.32). Critical to the formation of material **4.61** are the two strong hydrogen bonds formed between a geometrically matched amidinium cation (in **4.59**) and carboxylate anion (in **4.60**). Four anion binding units are appended to a tetrahedral receptor and then linked through the recognition of dicarboxylate anions to form an infinite assembly that crystallizes from aqueous solvent. The anion selectivity of the amidinium host for carboxylate guests means that no other anions were able to form this framework. Interestingly, it was found the pores of this framework can be used to encapsulate enzymes and extend the activity of these catalysts beyond that found under biological conditions.

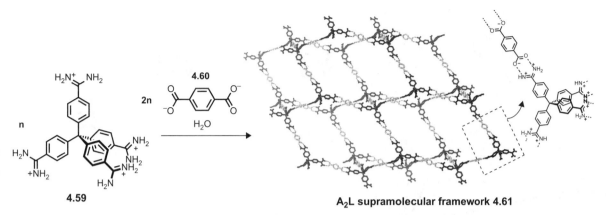

4.60

4.59

A₂L supramolecular framework 4.61

Scheme 4.32 Self-assembly of a supramolecular organic framework through specific hydrogen bond interactions between tetra-amidinium and dicarboxylate building blocks. Crystal structure graphic made by the original author. Refcode and source of first publication: CSD-1523338: M. Morshedi *et. al.*, *Chem. Sci.*, 2017, **8**, 3019.

4.5 Further applications of self-assembly

Self-assembly has enabled the construction of many elaborate and exotic molecular architectures but, beyond their aesthetics, we have also highlighted where systems are providing new functionality. From antibacterial agents and enzyme catalysis, to energy storage and chemical purification, self-assembly is being put to use in a diverse range of fields. Inevitably some of these applications, in particular those connected to materials chemistry, are at a more advanced stage than others. Integral to their purpose are the length scales at which these self-assembled systems are designed to operate. In this section, we have picked out a selection of molecular, macromolecular, and framework assemblies to showcase further exciting applications of this area of supramolecular chemistry.

Metal coordination cage **4.62** has been used to catalyse a Diels–Alder reaction between benzoquinone and a diene (Scheme 4.33). The three-dimensional cavity of this M_2L_4 capsule acts like an enzyme because the benzoquinone guest is activated for reaction upon its binding by hydrogen bond interactions. The Diels–Alder reaction therefore proceeds over 1000 times faster in the presence of the metal assembly than without. Furthermore, since the coordination cage is only selective for the starting material, it does not bind the product, ensuring low product inhibition.

Metal coordination cages that exhibit strong guest binding can function as phase transfer transporters for their molecular cargo (Scheme 4.34). By exchanging the hydrophilic sulfate counteranions of tetrahedral capsule **4.63** for hydrophobic tetrafluoroborate anions, the cage and its encapsulated fluoroadamantane guest move from the water to an ionic liquid layer of a biphasic solution. In the future, these cages could be used as low cost alternatives to discriminate, recycle, and separate precious hydrocarbons in place of energy intensive processes, such as fractional distillation.

With the application of biological substrate transport in mind, amphiphilic polymers containing discrete hydrophobic and hydrophilic sections (block copolymers) have been synthesized and assembled into spherical micelles of uniform size and shape in water (Scheme 4.35). To prepare these macromolecules for assessing targeted delivery, a fluorophore was appended to the hydrophobic block of the polymer scaffold, enabling the state of self-assembly to be reported through an optical response. Fluorescence spectroscopy revealed a switching on of the emission upon self-assembly as the fluorophore is embedded in the hydrophobic core of micelle

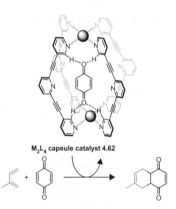

M₂L₄ capsule catalyst 4.62

Scheme 4.33 A self-assembled artificial Diels–Alderase.

M$_4$L$_4$ tetrahedron 4.63

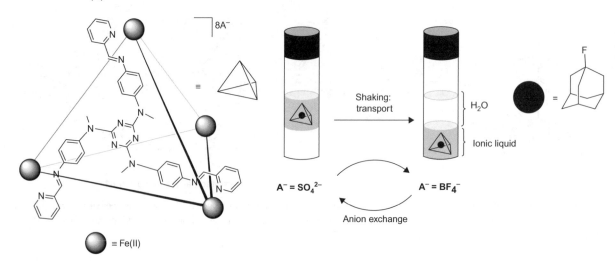

Scheme 4.34 Transport of molecular cargo courtesy of a M$_4$L$_4$ tetrahedral cage.

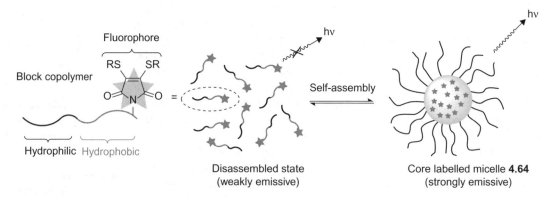

Scheme 4.35 The addition of a fluorophore to an amphiphilic polymer allows it to optically report on the state of a self-assembled micelle.

4.64, protecting it from collisional quenching with the solvent. Hence, this response can be employed to conveniently detect micelle ↔ molecule assembly–disassembly and hence the capture or release of molecular cargo.

The transport of substrates across cell membranes is one mechanism by which extracellular chemical signals can mediate intracellular functions. Alternatively, specific biological substrates that bind to protein receptors in pores of the cell membrane can also enable communication across the barrier. It has been shown that tetrahedral M$_4$L$_6$ cages (**4.12**) mimic these transmembrane messengers via enantioselective binding to these proteins (Figure 4.18). This is possible because each of the metal centres in a metallo-supramolecular assembly can have two possible chiral arrangements of the ligands (denoted Δ or Λ to distinguish axial chirality). Typically, only homochiral species are formed to give a racemic solution and thus, for cage **4.12**, Δ Δ Δ Δ and Λ Λ Λ Λ exist in equal amounts. Remarkably, protein receptors embedded in a lipid bilayer were able to interact with and discriminate between these chiral supramolecular cages. This indicates the potential of three-dimensional self-assembled cages to propagate a specific response to the interior of a cell.

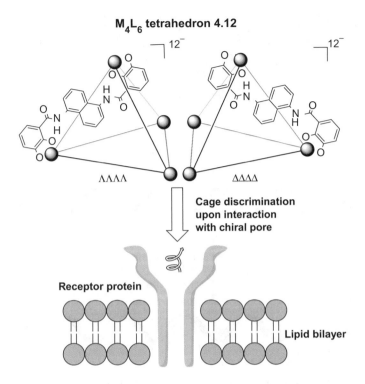

Figure 4.18 Discrimination between chiral supramolecular cages has been demonstrated using a protein nanopore.

Foldamers that form helical structures have inherent axial chirality since the helix has a screw sense that turns either to the right or to the left. A hydrogen bonded foldamer (**4.65**) has been developed in which the chirality can be reversibly controlled (Scheme 4.36). This system comprises a polyamide oligomer of 2-aminoisobutyric acid and a boronic ester installed at the N-terminus. In this state, there is no preference for either screw sense, and so a racemic mixture exists in solution. However, the binding of a chiral diol ligand at the boron receptor site induces a left-handed screw sense preference in the helix because of an intimate connection between the complex and the intramolecular network of hydrogen bonds. The addition of a stronger binding diol of opposite chirality triggered this screw sense to be switched. Therefore, this dynamic foldamer behaves like a transmembrane signalling protein in a cell membrane; reversible binding of a specific messenger molecule triggers conformational change, which in turn promotes the propagation of information (in this case chirality) over long distances.

Of the current self-assembled molecular systems, porous supramolecular organic frameworks represent some of the most developed in terms of functional supramolecular materials. In particular, the motivation for maximizing porosity of MOFs is driven by their applications in green energy, predominantly as gas storage materials. However, the binding of functional guests within MOF cavities is also important for sensing, biomedical imaging, and catalysis. An example of the latter is shown in Scheme 4.37 in which Crabtree's catalyst, an Ir(I) metal complex employed in the hydrogenation of alkenes, is encapsulated within Cr(III)-based MOF **4.66**. Normally this

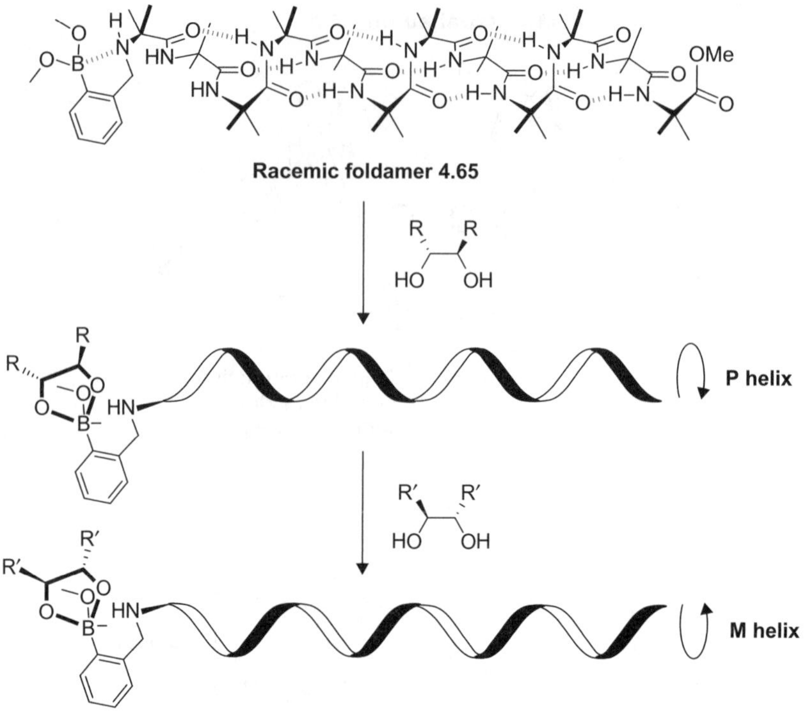

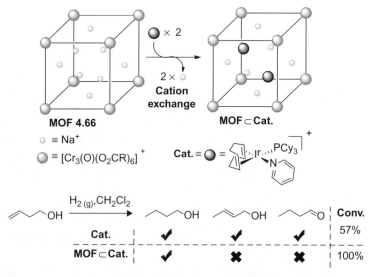

Racemic foldamer 4.65

Scheme 4.36 A foldamer acts as a biomimetic communicator through guest-induced changes in its axial chirality.

MOF 4.66

○ ≡ Na⁺

◉ ≡ [Cr₃(O)(O₂CR)₆]⁺

MOF ⊂ Cat.

Cat. ≡ ● ≡

					Conv.
Cat.	✔	✔	✔		57%
MOF ⊂ Cat.	✔	✘	✘		100%

Scheme 4.37 The loading of a catalyst within a MOF improves product conversion and selectivity in a hydrogenation reaction. The secondary building unit is based on Cr(III) and has a similar structure to that shown in Scheme 4.12. Figure adapted under the terms of the Creative Commons Attribution License from A. Grigoropoulos *et. al.*, *Angew. Chem. Int. Ed.*, 2018, **57**, 4532.

homogeneous catalyst is deactivated over the course of a reaction because coordination of hydride ligands leads to the formation of polymetallic clusters. However, by isolating the catalyst within the well-defined, hydrophilic environment of the MOF pores, the activity is enhanced and also imparts selectivity for hydrogenation of olefinic alcohols over the less desirable isomerization reaction.

4.6 Summary and conclusions

Self-assembly is the reversible association of molecules and/or ions to form larger and more complex supramolecular structures. The fact that each building block (neutral or charged) has been pre-programmed with the chemical information needed for this process to occur spontaneously is key. Therefore, synthetic effort is focused on the rational design of these building blocks, e.g. the preparation of ligands with the required donor atom geometry, since, in theory, the desired structure will be formed relatively easily as a single, stable structure in the final convergent step. This is providing access to structures that are beyond reach through conventional step-wise synthetic strategies.

Throughout this chapter we have seen examples of different variations of self-assembly such as strict self-assembly (e.g. complementary D–A hydrogen bond pairs), directed self-assembly (e.g. by discrete metal cations or inorganic anions) and assisted self-assembly (e.g. via π–π aromatic stacking or guest binding). There can be no doubt that our greater understanding and ability to harness the power of self-assembly is allowing the construction of ever more challenging and aesthetically pleasing molecular architectures. A diverse array have been described above including macrocycles, cubes, grids, helicates, cages, capsules, nanorings, porous frameworks, fibres, and tubes, but there are many more to be explored through the resources selected in the Further reading, Section 4.7.

It is exciting that many of these supramolecular structures are now finding worthwhile applications, and being designed with them in mind, across a wide range of fields. In particular, the ability to control a defined nanospace across multiple length scales is behind the success of guest capture by porous host assemblies for catalysis, purification, and energy storage.

Finally, self-assembly is of critical importance to the areas of biology, biochemistry, and medicine. Not only have natural systems provided inspiration for much of the work in this area, but a greater understanding of the structures and mechanisms of biological assemblies such as cell membranes, vesicles, enzymes, DNA, and viruses will certainly assist the development of next-generation healthcare and future medicines.

4.7 Further reading

The following resources provide further examples and applications of self-assembly.

- A wide range of discrete metallo-supramolecular architectures, templation strategies, and applications are described in these comprehensive review articles: *Chem. Soc. Rev.*, 2013, **42**, 1728–54 and *Chem. Rev.*, 2011, **111**, 6810–918. For a summary of metal coordination materials including polymeric and metal–organic frameworks see: *Chem. Rev.*, 2013, **113**, 734–77. A simple laboratory practical that enables the preparation of an Fe(II) triple helicate is described here: *J. Chem. Educ.*, 2018, **95**, 648–51.

- Further details of the thermodynamics of the hydrophobic effect, a key tool for self-assembly in water, can be found in this review article: *Chem. Soc. Rev.*, 2015, **44**, 547–85. Investigations into the thermodynamics of other non-covalent interactions involved in self-assembly, based on the system outlined in Figure 4.16, are provided in *Chem. Soc. Rev.*, 2007, **36**, 172–88. DNA origami was first reported in this publication: *Nature*, 2006, **440**, 297–302.

- A number of additional discrete anion templated assemblies are described in this review: *Org. Chem. Front.*, 2018, **5**, 662–90. Further details of the anion framework materials containing polyhedra of the kind seen in Figure 4.17 are summarized in this article: *Acc. Chem. Res.*, 2016, **49**, 2669–79.

4.8 Exercises

4.1 a) Consider the ligand **Lₐ** below. How many metal cation binding domains does it have? For what reason might you expect it to self-assemble into a grid architecture over a helicate in the presence of Ag(I) cations?

b) Sketch out the structure of a hypothetical 5 × 5 ligand molecular grid and predict the form of the ^{109}Ag NMR spectrum it would generate.

c) Experimental observations indicate the only grids formed by self-assembly of **Lₐ** and Ag(I) measure 2 × 5 ligands. Sketch their possible structure and comment on how their ^{109}Ag NMR spectrum will differ to that from part (b). Suggest a possible reason for this outcome.

d) What synthetic strategy is used to prepare the structures above and ladder assembly **4.19** shown in Scheme 4.9? What architecture might be possible if the 'rungs' of **4.19** were swapped for ligand **L_b** below? (Hint: how many metal cation binding domains does **L_b** have?)

4.2 a) Considering the principle of maximum occupancy and the coordination geometry preferences of copper ions, predict the outcomes of these self-assembly reactions (note the different reaction conditions). Are any of the products chiral and if so why?

i)

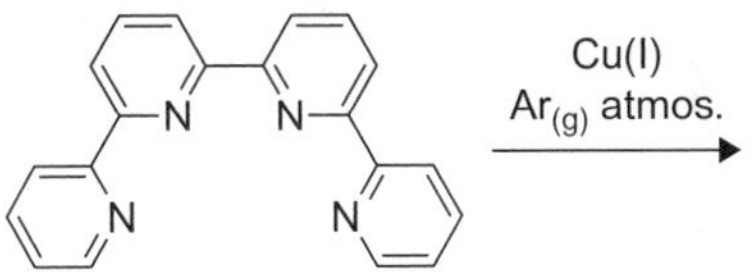

Cu(I)
Ar$_{(g)}$ atmos.

ii)

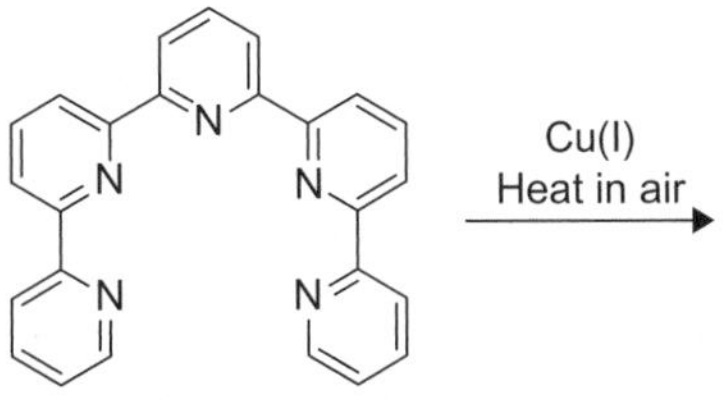

Cu(I)
Heat in air

b) How many different M$_3$L$_2$ cage compounds are possible from the self-assembly of ligands **L$_c$** and **L$_d$** and Pd(II) complex **M** shown below? How might one bias the formation of one cage over the others? (Hint: the nitrate ligands are labile.)

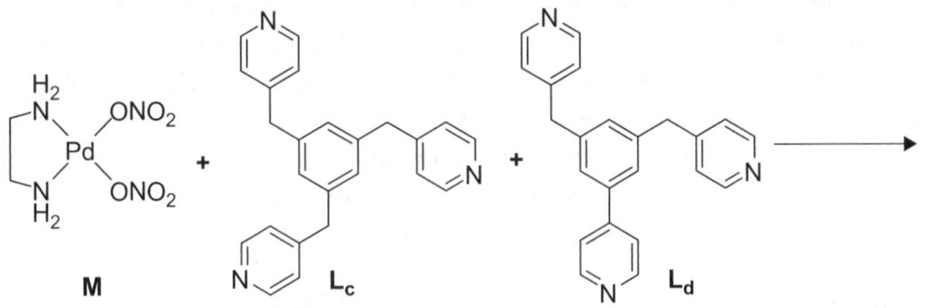

M   **L$_c$**   **L$_d$**

c) Design an alternative Vernier template for the synthesis of nanoring **4.18** from the same starting materials shown in Scheme 4.8.

4.3 a) Scheme 4.16 shows how structural modifications to cyanuric acid and melamine can alter the outcome of hydrogen bond self-assembly to generate a single rosette (**4.32**) instead of an insoluble polymeric network (**4.31**). Suggest how the hydrogen bond donor and acceptor patterns of these building blocks could be altered to form a one-dimensional polymeric chain as well. How might the length of these chains be altered and what effects would this have on the physical properties of the solution? Consider the effects of monomer concentration in your answer too.

b) In light of the solvent dependency of π–π aromatic stacking interactions, why do you think that toluene was chosen as the solvent for generating supramolecular polymer **4.43** in Scheme 4.21? What other solvents might be appropriate for polymerization? Include in your answer the possible implications of switching to a polar protic solvent.

For the answers to these exercises, visit the online resources which accompany this primer.

Mechanically interlocked molecules

5

5.1 Introduction

Mechanically interlocked molecules (MIMs) belong to a fascinating category of structures containing two or more molecular sub-components that, despite not being chemically bonded, are inextricably connected due to the inability of bonds to pass through one another. This type of molecular entanglement is described as a **mechanical bond** and, by fulfilling the criteria of Pauling's definition of a chemical bond, is therefore the newest type of bond in Chemistry! Mechanical bonding is what separates MIMs from the host–guest complexes and non-covalent assemblies presented in Chapters 1 to 4; these are molecules in their own right and not supramolecular entities. However, since templating intermolecular interactions were used to produce the first MIMs in significant quantities, their origins are deep-rooted in supramolecular chemistry.

The two archetypal MIMs are **catenanes** and **rotaxanes** (Figure 5.1). A catenane, derived from the Latin word for chain (catena), is a MIM containing two or more linked macrocyclic rings. A rotaxane is comprised of a macrocyclic wheel (rota) threaded by an axle (axis) molecule in the shape of a dumbbell, where the sterically bulky groups at either end prevent the ring from dethreading. The '[n]' prefix used in MIM nomenclature indicates the number of component parts, for example a [2]catenane consists of two interlocked rings (Figure 5.1). Mechanical bonds are distinct from conventional chemical bonds because they are only maintained by the physical forces that prevent bonds from passing through one another. Hence, in terms of their relative strength, a mechanical bond is as strong as the weakest chemical bond, typically a covalent one, in one of the MIM's constituent parts. For example, breaking open the macrocycle components of either a catenane or a rotaxane will also sever the mechanical bond, allowing the molecular sub-units to come apart.

Mechanical bonds are, in fact, very familiar from the macroscopic world; apart from the similarities between catenanes and the links of a chain, the washer encircling the thread of a screw and the wheels on the axle of a bicycle both resemble the interlocked architecture of a rotaxane (Figure 5.2). These analogies also highlight another important feature of mechanical bonds, namely that their components can be static, like those of a tightened screw, or dynamic, like those of a spinning bicycle wheel. On the atomic scale, functional MIMs are not only limited to synthetic molecules from the laboratory; indeed Nature frequently employs mechanical bonds in both mobile and stationary scenarios. For example, a rotaxane architecture is adopted during the

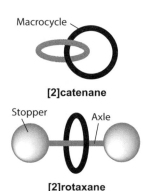

Macrocycle

[2]catenane

Stopper   Axle

[2]rotaxane

Figure 5.1 A [2]catenane and a [2]rotaxane, archetypal mechanically interlocked molecules.

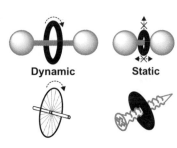

Dynamic   Static

Figure 5.2 Mechanically interlocked components can be dynamic or static.

biosynthesis of polypeptides; a chain of messenger RNA (axle) is clamped by a ribosome (macrocycle) as it moves between codons and transcribes it into a sequence of amino acids. Furthermore, the anchoring of an enzyme's peptide sub-units via their interlocking in the **chemical topology** of a catenane has been shown to increase the stability of its macromolecular structure. The examples of artificial rotaxanes and catenanes highlighted in this chapter also demonstrate new applications arising from static or dynamic MIM properties (Figure 5.2), including enantioselective catalysis and **molecular machines**.

At this point, it would be helpful to clarify some further terminology related to mechanically interlocked molecules. An important difference between a rotaxane and catenane is that only the latter is considered to have a non-trivial chemical topology (Figure 5.3). This means that to break the mechanical bond arising from a molecular link in a catenane also requires the breaking of a chemical bond. Containing two crossing points, a [2]catenane is in fact the simplest member of a larger topological family of 'prime links'. By contrast, the chemical topology of a rotaxane is relatively simpler because it is conceivable that the macrocycle and axle components may be separated by their continuous deformation, for example, by stretching the ring so that it can pass over one of the bulky stoppering units. Therefore, while both catenanes and rotaxanes are described as MIMs because they contain one (or more) mechanical bonds, it is strictly correct to refer to 'topology' and 'architecture' respectively when discussing the spatial arrangements of their molecular frameworks. Another category of topologically complex molecules are knots, although these structures do not contain mechanical bonds because they consist of a single molecular component that has become entangled within itself. Finally, it should be noted that there are a number of exotic MIMs (e.g. rotacatenanes) that resemble both catenanes and rotaxanes, demonstrating that this is an evolving field of research. Beyond the scope of this chapter, further details of these MIMs can be found via the resources in Section 5.6.

This chapter begins by outlining the formidable challenges presented in the syntheses of MIMs and the key strategies that have been developed to overcome them. The resulting preparation of rotaxanes and catenanes in high yields has enabled this field to flourish, making explorations into the properties and applications of MIMs an active area of modern-day chemical research. This is the focus of the second half of this chapter; starting with cases in which the influence of the mechanical bond on chemical and physical properties, including supramolecular properties, delivers applications in sensing, catalysis and materials. The control of molecular motion is one of the most prominent attributes of the mechanical bond; for example, the rings of a [2]catenane are able to move independently within the constraints of each other. The final sections of this chapter outline how this facet serves as the foundation of modern-day molecular machinery and how dynamic MIMs are critical to future applications in nanotechnology.

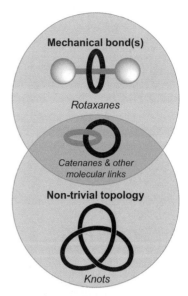

Figure 5.3 A Venn diagram to illustrate the main types of mechanically interlocked molecule and associated terminology.

5.2 Synthesis of mechanically interlocked molecules

The term 'mechanical bond' was first used in the 1950s when chemists were considering the possibility of molecules composed of linked rings, such as a [2]catenane. The next decade saw the first synthetic reports of catenanes, but these were plagued by extremely low yields due to their reliance on the serendipitous interpenetration of macrocycles and their linear precursors. It was not until twenty years later, with the establishment of supramolecular host–guest chemistry, the development of rationally designed passive and, most recently, active metal templates, that mechanically inter-locked molecules could be prepared in appreciable yields. While today rotaxanes and catenanes may be considered as more mainstream molecular targets, the synthetic challenge presented by forming mechanical bonds in high yields continues to inspire new approaches and strategies, including the preparation of MIMs with multiple in-terlocked molecular components, known as 'higher-order' architectures.

The primary synthetic strategies used for making rotaxanes and catenanes are out-lined in Figure 5.4, with the most common being 'stoppering' and 'clipping' approaches respectively. These both proceed via the ubiquitous ***pseudorotaxane*** intermediate, an interpenetrated, but not interlocked, assembly formed by the threading of a lin-ear molecule through a macrocyclic ring. This association is driven by intermolecular non-covalent interactions, highlighting the importance of supramolecular host–guest chemistry in developing robust pseudorotaxane assemblies to allow the capture of mechanical bonds in high yields. The larger number of synthetic routes to rotaxanes over catenanes is a consequence of their difference in topological complexity; whilst rotaxanes may also be prepared by swelling, shrinkage, or slippage strategies, no man-ner of stretching, bending, or rotating of the covalent bonds in macrocyclic rings will deliver catenane analogues. For these, the formation of a mechanical bond is con-comitant with the formation of one or more covalent bonds.

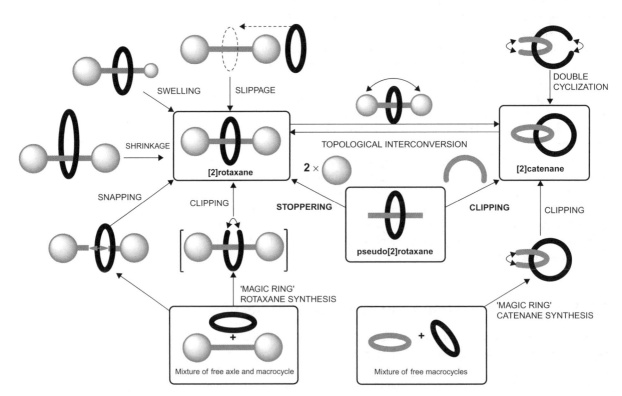

Figure 5.4 The primary synthetic strategies for rotaxanes and catenanes.

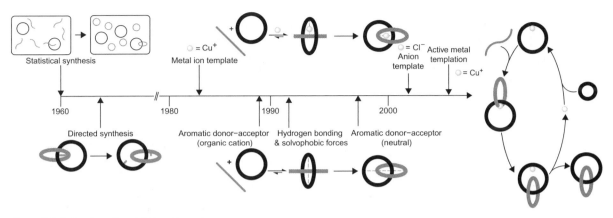

Figure 5.5 A timeline of key MIM syntheses.

In the following sections, the milestones in MIM syntheses are presented in chronological order alongside some more recent poignant examples. These key breakthroughs and the synthetic approach adopted have been illustrated in Figure 5.5. Starting out from Frisch and Wasserman's statistical synthesis (1960) and Schill's directed synthesis (1964), it was Sauvage's use of a Cu(I) metal ion template to direct the synthesis of a [2]catenane (1983) that first enabled MIMs to be prepared in appreciable yields. Stoddart's introduction of aromatic donor–acceptor interactions as templates for MIMs (1989) was shortly followed by Hunter and Vögtle's exploitation of hydrogen bonding in the synthesis of [2]catenanes (1992). At the same time, solvo-phobic forces were employed by Harada in the synthesis of poly[n]rotaxanes composed of cyclodextrins (1992). While Sanders demonstrated that aromatic stacking interactions between neutral aromatic motifs could be harnessed for MIM synthesis (1997), Loeb furthered the use of charge transfer interactions involving organic cations to prepare [2]rotaxanes in high yields (1998). A few years later, the library of discrete ionic templates for MIM syntheses was expanded to include anions by Vögtle (pheno-late) and Beer (2002, halides) and further transition metal complexes by Leigh (2003). Out of the latter sprang Leigh's active metal templation strategy (2006), in which Cu(I) acts as both a template for directing the assembly of MIM precursors and a catalyst for the covalent bond-forming reaction that captures the mechanical bond.

Statistical and directed syntheses

The first generation of MIMs made use of the serendipitous threading of a macrocyclic ring by a linear molecule, to form an interpenetrated assembly as the precursor to a mechanical bond. In 1960, Frisch and Wasserman described the clipping reaction of a diester by an acyloin condensation that, when statistically threaded through a cycloal-kane, forms a MIM composed of two interlocked rings (**5.1**). They coined this molecule a catenane (Scheme 5.1). When finally separated from starting materials and side-products, the [2]catenane yield was found to be extremely low (< 1%). The same was true for the synthesis of [2]rotaxane **5.2** (6%). In both cases, an absence of associative non-covalent interactions between the two components is responsible for the low probability of interpenetration.

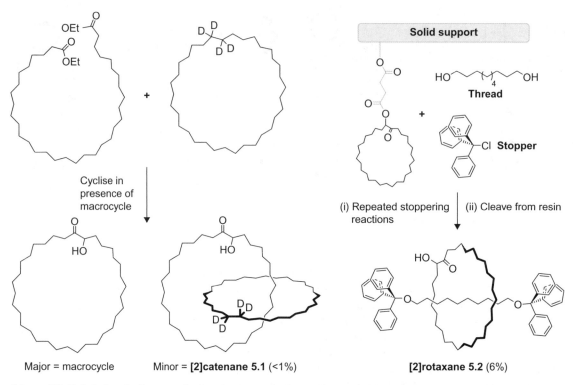

Scheme 5.1 Statistical synthetic approaches to catenanes and rotaxanes.

An early strategy to address this problem was Schill's use of a cleavable covalent bond to initially connect two ring components and then break it to generate a [2]catenane (**5.3**) in the final step of the synthesis (Scheme 5.2). The mechanical bond-forming step of this synthesis is indeed nearly quantitative; however, the overall yield of MIM remains low owing to the large number of steps needed to construct the [2] catenane precursor.

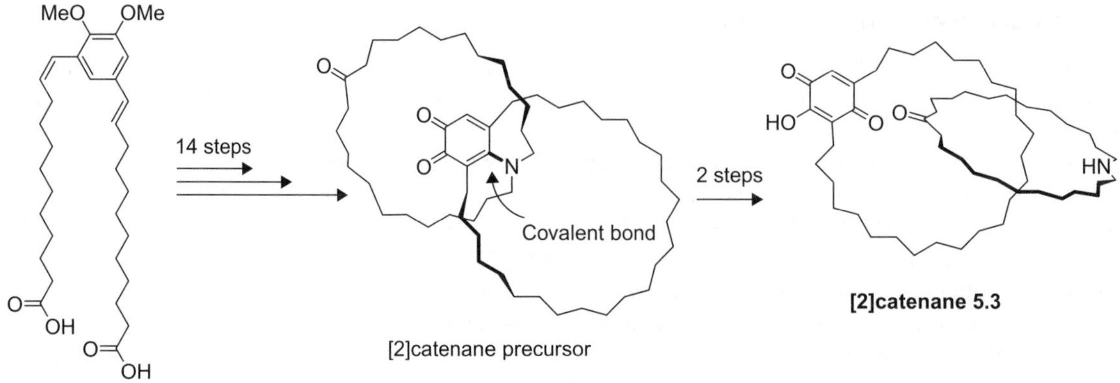

Scheme 5.2 An early example of the directed synthesis of a [2]catenane.

Plagued by difficulties in low yields and product separation, it was apparent that a key challenge in MIM construction arose from the difficulty in arranging two or more independent molecules to maximize their probability of forming a mechanical bond. Taking inspiration from supramolecular chemistry, the use of molecular recognition and templation led to a paradigm shift in the high-yielding syntheses of catenanes and rotaxanes.

Metal template-directed syntheses

As we have seen in Chapter 4, a template is any chemical species that can orientate and direct reactants to favour the formation of a particular product. Metal ions make for efficacious templates; indeed, strong metal–ligand coordinate bonding is responsible for directing the self-assembly of many of the macrocycles, cages, and frameworks described in Section 4.2. Pioneered by Nobel laureate Jean-Pierre Sauvage, it was the advent of transition metal cation templation that also revolutionized the synthesis of mechanical bonds in appreciable yields. With the aim of developing new photocatalysts

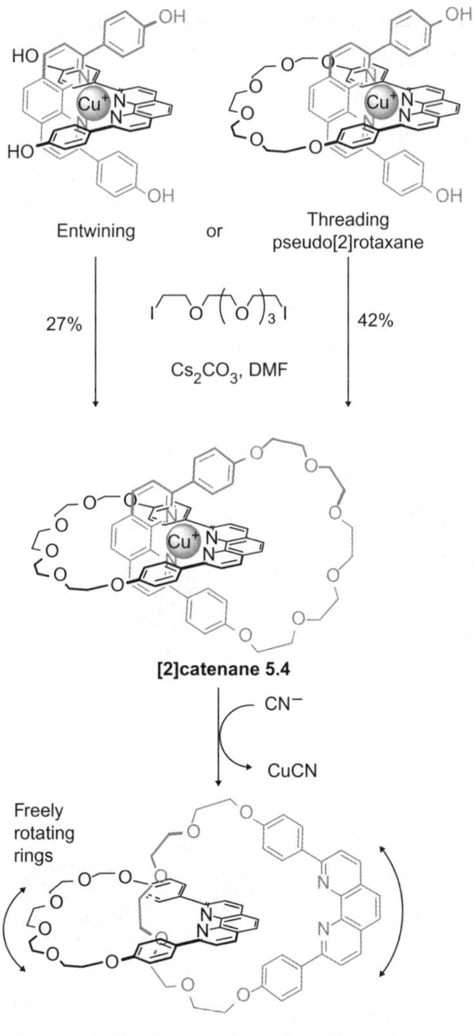

Scheme 5.3 The first metal template-directed approach for MIM synthesis used Cu(I) chelation.

for water oxidation, Sauvage was experimenting with a Cu(I) metal ion to arrange two phenanthroline ligands in an orthogonal fashion. Critically, he realized that this entwined geometry was appropriate for forming a [2]catenane upon macrocyclization of each ligand set. This was subsequently performed by Williamson ether synthesis to deliver MIM **5.4** in a landmark yield of 27% (Scheme 5.3). In his seminal report, Sauvage also outlined a second synthetic approach in which the metal cation directs the threading of one phenanthroline ligand through the macrocycle of another before its clipping to form the same [2]catenane but in even higher yield (42%). Alongside ^{1}H NMR spectroscopy and X-ray diffraction structural analysis, the mechanical interlocking of the two rings was demonstrated by mass spectrometry. The cleavage of a covalent bond simply results in macrocycle de-threading; therefore, no species was detected between the m/z of a single macrocycle and that of the catenane molecular ion peak.

Importantly, the Cu(I) template could be removed from the [2]catenane but required an excess of strongly coordinating CN⁻ ligands due to the mechanical bond-enhanced stability of the metal ion complex. This generated the 'free' catenane ligand with a unique interlocked cavity and dynamic properties from rotating rings. Both these findings provide the basis for exploiting the mechanical bond for molecular recognition and molecular machine applications (Section 5.4).

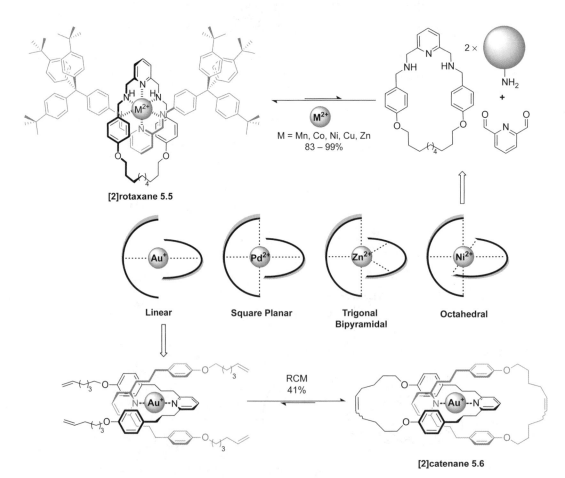

Figure 5.6 Common transition metal coordination geometries for MIM precursor assemblies, including linear Au(I), used for preparing a [2]catenane. (RCM = ring-closing metathesis.)

The discovery of template-directed synthesis of MIMs by metal cations represented a huge leap forward in forming mechanical bonds. Tolerant to a wide range of reaction conditions, robust metal–ligand complexes have been designed to possess the requisite crossing points or entanglements to prepare a wide variety of functional rotaxanes and catenanes. In analogy to metal-directed self-assembly (Section 4.2), the stereochemical preference of the interweaving metal template and the principle of maximum occupancy (to satisfy all coordination sites, see Section 4.2), provide powerful tools to assemble and orientate components and define the topology of the MIM product. A range of transition metal cation coordination geometries e.g. linear Au(I), square planar Pd(II), and trigonal bipyramidal Zn(II), have been employed in the synthesis of MIMs (Figure 5.6). In particular, excellent yields have been reported for a range of octahedral coordinated metal(II)-templated [2]rotaxanes, constructed under thermodynamic control using Schiff base chemistry. For example, MIM **5.5** is formed in a five component self-assembly process in a one-pot synthesis without the need for additional purification (Figure 5.6). Furthermore, Au(I) metal ions have been used to preform two monodentate pyridyl ligands in a linear geometry (N–Au–N bond angle = 175°). The [2]catenane **5.6** was isolated in 41% yield upon a double ring-closing metathesis (RCM) cyclization of these entwined macrocycle precursors (Figure 5.6). Alongside imine formation, olefin metathesis is a widely used reaction in thermodynamic mechanical bond synthesis because of its wide substrate scope and reversibility.

Transition metal-directed self-assembly has been harnessed to prepare architectures with even more complex topologies. One elegant example includes the network of Borromean rings formed upon crystallization of a tris-pyridyl functionalized cyclotriveratrylene ligand in the presence of Cu(II) cations (Scheme 5.4). Molecular Borromean rings (**5.7**) are composed of three independent macrocycles that are linked together, but where no two rings are interlocked in a catenane. Therefore, the breaking of any one of these rings will cause the link to come apart.

Alkali metal cations have also been exploited for templating the synthesis of MIMs. For example, Na⁺ can template the formation of pseudorotaxane assemblies between crown ether-like macrocycles and a range of amide or urea-based threading components. These are stabilized through cation–dipole and NH···O hydrogen bonding interactions between

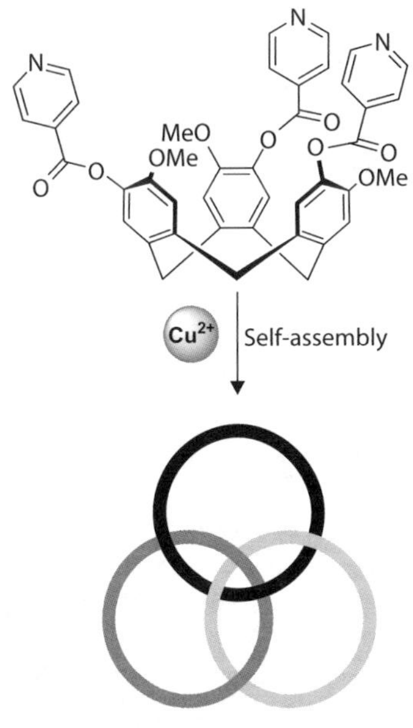

Borromean rings 5.7

Scheme 5.4 The Cu(II) directed self-assembly of a crystalline network of Borromean rings. This is illustrated here as a simpler Venn projection.

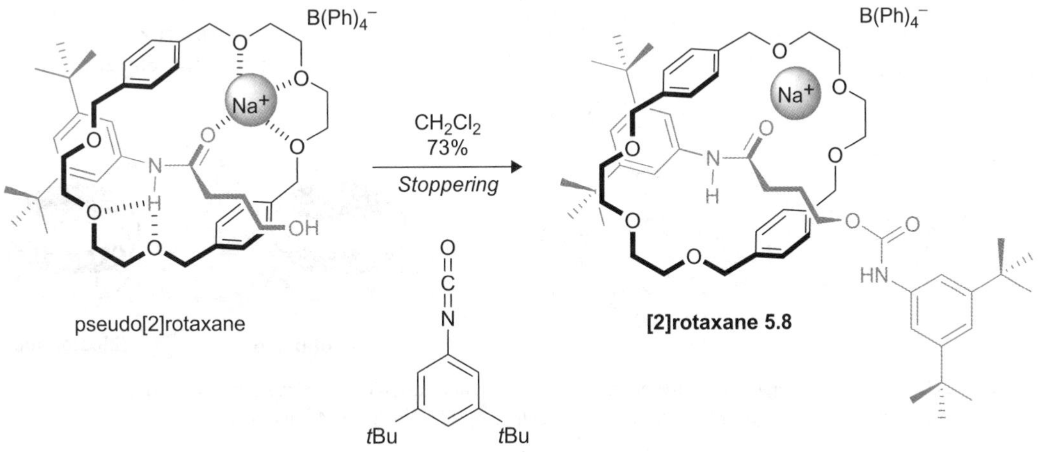

Scheme 5.5 A sodium ion-templated [2]rotaxane is formed via a mono-stoppering procedure.

electron donating crown ether oxygen and amide/urea protons. Stoppering of one set of these building blocks generated [2]rotaxane **5.8** in high yield (Scheme 5.5).

More recently, a synthetic approach has been developed in which the metal ion acts as both a template and a catalyst for the mechanical bond forming step, making it distinct from the 'passive' metal templates described above. Known as active metal templation (Figure 5.5), the metal ion preorganizes the MIM precursors and promotes their reaction to favour concomitant covalent bond and mechanical bond formation, thereby generating the rotaxane or catenane in high yield. This strategy is outlined in more detail later in this section.

Aromatic donor–acceptor template-directed syntheses

Highlighted throughout the earlier chapters, aromatic stacking interactions are important in supramolecular chemistry for guest recognition and directing self-assembly. In respect of MIM synthesis, the face-centred stacking of an electron-deficient organic cation π-acceptor with an electron rich π-donor also represents a popular templating assembly (Figure 5.7). Frequently these donor–acceptor (D–A) interactions (see Section 4.3) generate a new and easily identifiable band in the UV-vis absorption spectrum, providing a spectroscopic signature for the interpenetrated assembly and a handle for quantifying its strength, i.e. calculation of a K_a for the pseudorotaxane assembly.

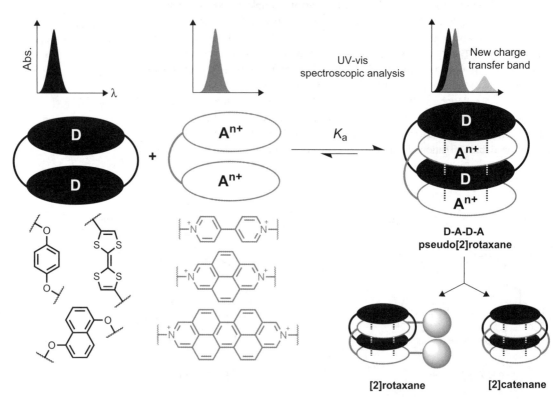

Figure 5.7 The charge transfer interactions between various π-donors and cationic π-acceptors can be used to prepare interpenetrated assemblies, as characterized by absorption spectroscopy. D = π-donor, A = π-acceptor.

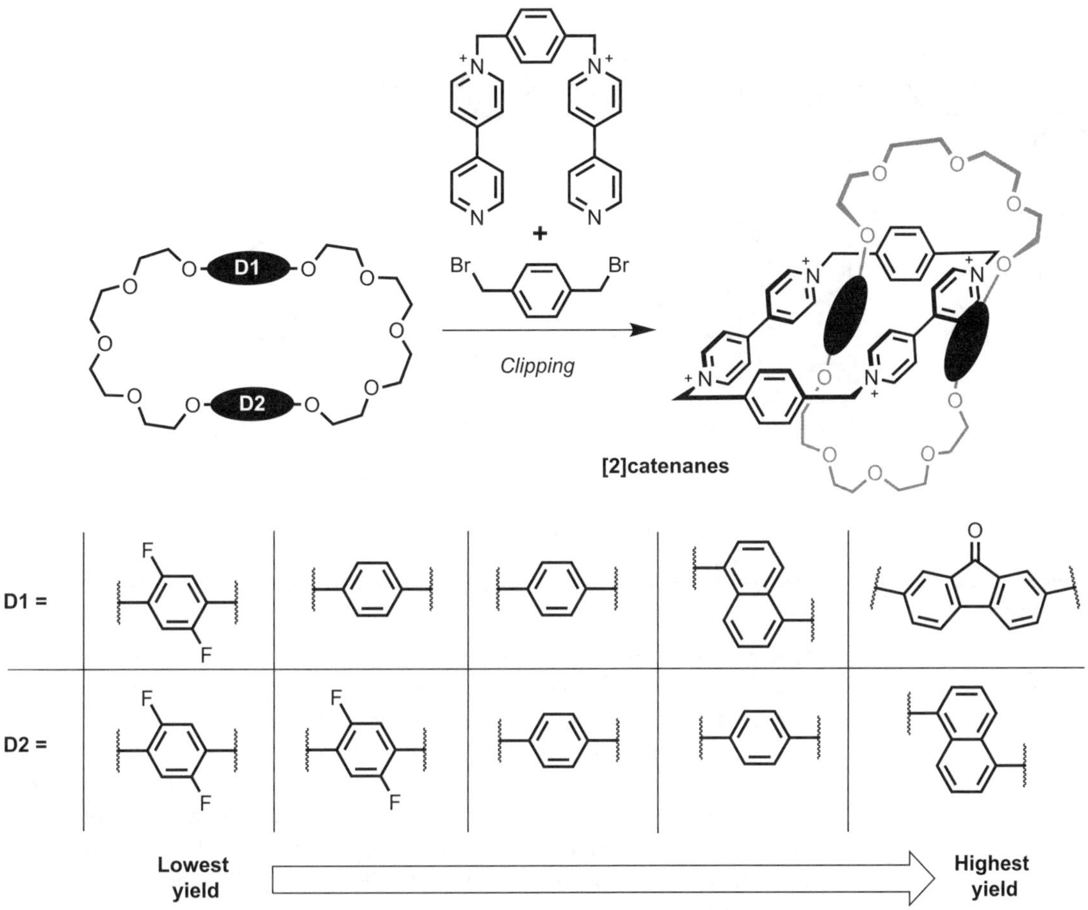

Scheme 5.6 The yields of 'blue box'-based [2]catenanes by clipping follows trends in donor strength and π surface area.

Planar aromatic dications such as bipyridinium (also known as viologen), diaza-pyrenium, and diazaperopyrenium are regularly employed as the π-acceptor components of interpenetrated assemblies (Figure 5.7). These can be paired with neutral π-donors such as hydroquinone, dialkoxynaphthalenes, tetrathiafulvalene, or free-base porphyrins. To maximize the strength of the D–A charge transfer interactions, pseudorotaxane assemblies are designed as sandwich complexes in which a π-donor/acceptor guest is bound between two aromatic motifs of a macrocyclic host. With electronic complementarity between their components, these D–A–D–A assemblies are common precursors to rotaxanes or catenanes.

Nobel laureate J. Fraser Stoddart reported a tetracationic macrocycle containing two 4,4′-bipyridinium units connected with xylyl linkers, nicknamed the 'blue box' on account of its colour upon reduction (**3.22**, Figure 3.23A). Introduced in Chapter 3 (Section 3.3), cyclophane **3.22** was shown to recognize π-electron rich guests inside its π-electron deficient cavity (Figure 3.23B), in both organic and

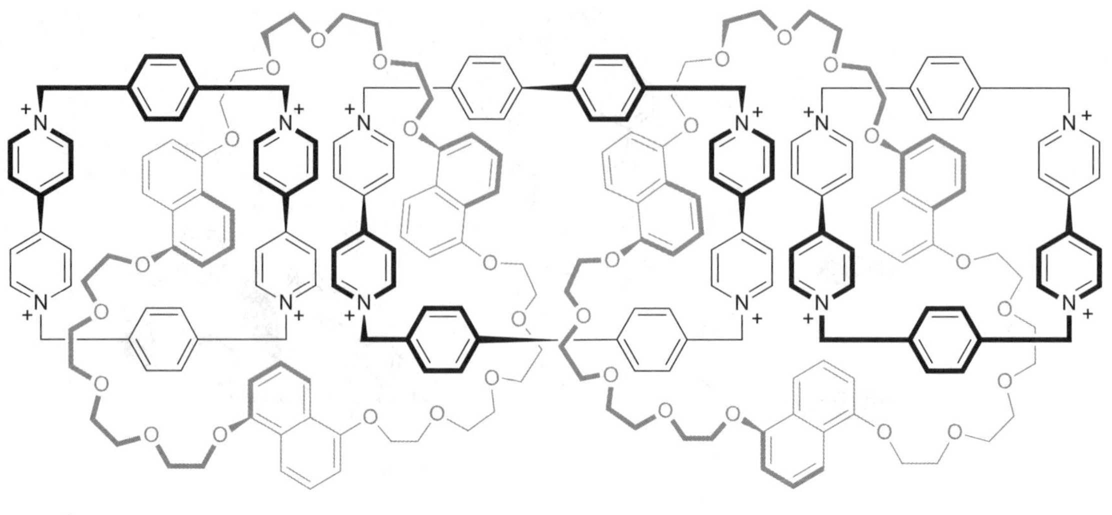

[5]catenane 5.9 = Olympiadane

Figure 5.8 The structure of [5]catenane olympiadane, the synthesis of which is templated by donor–acceptor charge transfer interactions.

aqueous media. Envisaging these robust organic cation-based assemblies as precursors to MIMs, both clipping and stoppering strategies have been successfully employed to construct a plethora of catenanes and rotaxanes containing **3.22**. In particular, extensive work on the preparation of [2]catenanes, via the clipping of the 'blue box' around a bis-π-donor-based macrocycle of electronic complementarity, has mapped out an important relationship between MIM yield and π–π interaction strength (Scheme 5.6). As a general rule, increasing the strength of the π-donors of one macrocyclic component maximizes the [2]catenane yield, as does increasing the size of the dioxyarene π system.

Further, structural modifications have seen the xylyl spacers of the blue box replaced with biphenyl, thereby doubling the width of the cyclic cavity and creating a tetracationic molecular square large enough to accommodate two π-donor-based macrocycles, forming a [3]catenane. This was exploited in the synthesis of [5]catenane **5.9**, named 'Olympiadane' due to its similarity to the five interlocked Olympic rings and achieved by interlocking two further blue box macrocycles (Figure 5.8).

More recently, it was discovered that the chemical reduction of bipyridinium units to their corresponding radical cations negated the need for D–A interactions of electronic complementarity, allowing the synthesis of an octacationic homo[2]catenane comprised of two 'blue box' macrocycles (Scheme 5.7). While bipyridinium will not thread inside the cavity of macrocycle **3.22** on the basis of electrostatic repulsion, the use of a reducing agent (Zn dust) under an inert atmosphere generates a tris-radical tetracationic pseudo[2]rotaxane **5.10** ($K_a = 10^5$ M^{-1}), a key intermediate in the preparation of [2]catenane. Intriguingly, upon exposure to air the bis-radical [2]catenane **5.11** remains in a reduced form. Whilst open-shell species are not typically air and water stable, this observation suggests that mechanical bonds, like supramolecular cages (Section 4.2), can stabilize reactive species. A strong chemical oxidant was required to convert **5.11** to its closed-shell redox state, **5.12**. Electrochemical studies on octacationic [2]catenane **5.12** revealed that up to eight electrons may be pumped into each molecule, highlighting its potential application as an energy storage material.

A combination of aromatic stacking interactions and C–H⋯O hydrogen bonds between the organic cation 1,2-bis(pyridinium)ethane and macrocycle dibenzo[24] crown-8 were used to assemble pseudo[2]rotaxane **5.13** (Scheme 5.8). This subsequently formed [2]rotaxane **5.14** upon stoppering of the thread. Interestingly, the D–A charge transfer interactions between π-donor rings of the host and the cationic guest are dominant in stabilizing the interpenetrated assembly since switching the dibenzo macrocycle for its parent crown ether (i.e. [24]crown-8) significantly reduces the association constant (K_a = 1,200 M^{-1} to K_a = 320 M^{-1} in acetonitrile). In contrast, the appendage of negatively charged sulfonate groups to the dibenzo[24]crown-8 macrocycle increases its affinity for the cationic guest, enabling the pseudo[2]rotaxane to be assembled in polar, protic solvents such as MeOH.

In organic solvents, it is also possible to exploit D–A aromatic stacking between neutral motifs to template the formation of mechanical bonds. For example, π-acceptor motifs such as naphthalene diimides and pyromellitic diimides form interpenetrated assemblies with a dioxyarene π-donor-based macrocycle. In DMF, Eglington–Glaser–Hay coupling can be used for a [2+2] cyclization of bis-alkyne derivatives containing either π-acceptor around the π-donor macrocycle, generating [2]catenanes **5.15** and **5.16** with the familiar D–A–D–A aromatic stacking pattern (Scheme 5.9). A greater yield is achieved with the larger π-acceptor (i.e. [2]catenane **5.15**) on account of stronger templating interactions.

Dynamic combinatorial chemistry has been used to explore the scope of catenane syntheses directed by neutral aromatic templates in aqueous media. A reversible disulfide exchange mechanism enables water soluble π-donor and π-acceptor macrocyclic precursors to assemble in various combinations to generate covalent bonds (macrocycles) and subsequently [2]catenanes (Scheme 5.10). Due to the aqueous conditions, the electrostatic contribution of aromatic donor–acceptor charge transfer interactions is now exceeded by hydrophobic forces, thereby explaining the formation of hetero[2]catenanes such as **5.17** that deviate from the familiar D–A–D–A aromatic stacking pattern. Indeed, it is even possible to prepare an A–A–A–A [2]catenane in the absence of π-donor starting materials. An advantage of this thermodynamic-controlled approach is its adaptability; the ratio of MIMs in solution can be adjusted by exposing the system to different stimuli. For example, while the introduction of bipyridinium as a π-acceptor favours D–A–D–A charge transfer interactions, the addition of salt augments the hydrophobic contribution to mechanical bond templation, triggering homodimerization.

An alternative thermodynamic-controlled approach has been used to prepare D–A [2]rotaxane **5.18**, in which a perylene diimide axle acts as a π-electron deficient recognition unit for a dioxynaphthalene-based macrocycle (Scheme 5.11). In this one-pot synthesis, the preformed, yet non-interlocked, axle and macrocycle starting materials are heated together for two weeks in a CHCl$_3$/MeOH solvent mixture. After this time, a proportion of the rings were found to have threaded onto the axle, as evidenced by new through-space coupling between interlocked components

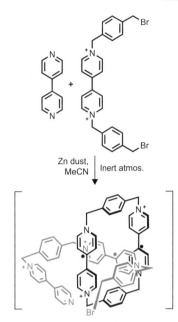

Zn dust, MeCN | Inert atmos.

pseudo[2]rotaxane 5.10

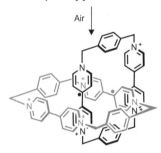

Air

open shell [2]catenane 5.11

Strong oxidant

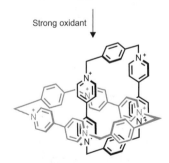

closed shell [2]catenane 5.12

Scheme 5.7 Chemical reduction enables the formation of an octacationic [2]catenane via mechanical bond-stabilized radicals.

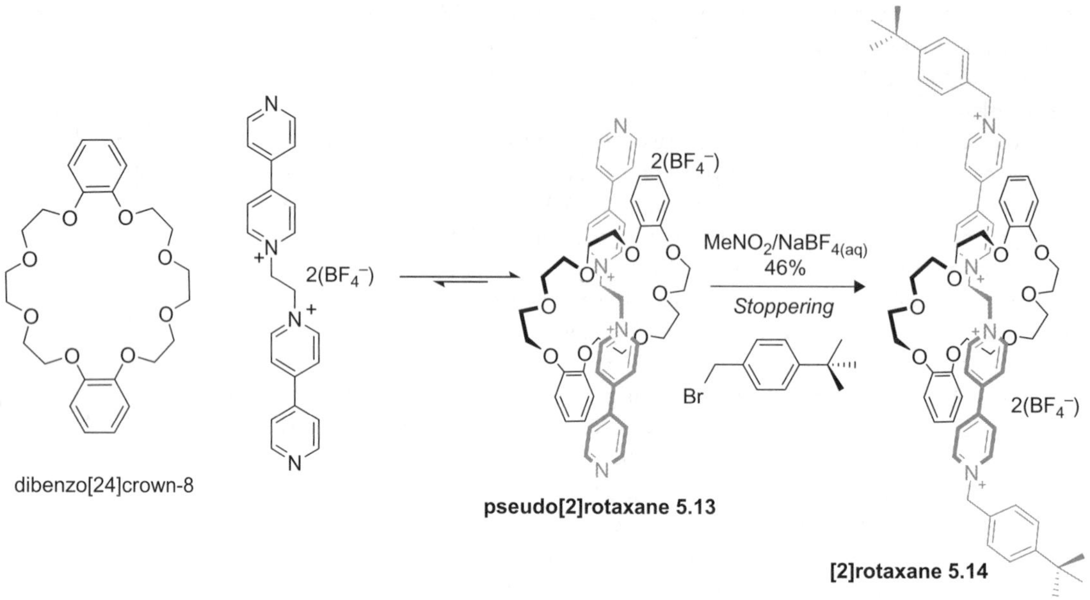

dibenzo[24]crown-8

2(BF$_4^-$)

2(BF$_4^-$)

MeNO$_2$/NaBF$_{4(aq)}$
46%

Stoppering

Br

pseudo[2]rotaxane 5.13

2(BF$_4^-$)

[2]rotaxane 5.14

Scheme 5.8 A [2]rotaxane is prepared by stoppering a dibenzo[24]crown-8–bis(pyridinium)ethane pseudo[2]rotaxane assembly.

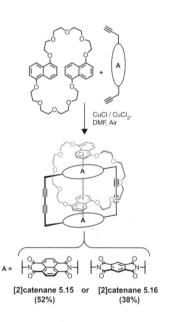

CuCl / CuCl$_2$,
DMF, Air

A =

[2]catenane 5.15 or [2]catenane 5.16
(52%)   (38%)

Scheme 5.9 Synthesis of the first [2]catenanes containing neutral aromatic motifs. A = π-acceptor.

in the two-dimensional NMR spectrum. Known as 'slippage', this strategy relies on the fact that at room temperature the activation barrier for threading is insurmountable but, upon heating, macrocycles gain enough energy to 'slip over' the bulky stopper group, generating a pseudo[2]rotaxane assembly. Importantly, upon cooling, the interpenetrated macrocycles become trapped on the axle component, enabling isolation of a [2]rotaxane (**5.18**). This highlights a subtle difference in the mechanical bonding between catenanes and rotaxanes. While the former requires a covalent bond to be broken in order to sever the topological link, the latter can be considered a 'kinetically stable pseudorotaxane' because it is dependent on the activation barrier to de-threading. This is further demonstrated by [2]rotaxane **5.18** because electrochemical reduction of the perylene diimide weakens the D–A interactions, triggering partial dissociation of the MIM components.

More recently, an ambitious class of mechanically bonded architectures have been reported that resemble both rotaxanes and pseudorotaxanes. A carbon nanotube was used as an axle component by clipping a macrocycle precursor containing two pyrene panels around the nanotube cylinder using ring closing metathesis, templated by aromatic stacking interactions (Scheme 5.12). A series of washes and filtrations are required to isolate the mechanically interlocked nanotubes (**5.19**) which, due to their size, can be characterized by transmission electron microscopy. It was found that stopper groups are not required to prevent these rings from slipping off the nanotube. Indeed, the mechanical bonds cannot be broken without heating to excessively high temperatures which indicates these architectures are best described as rotaxanes.

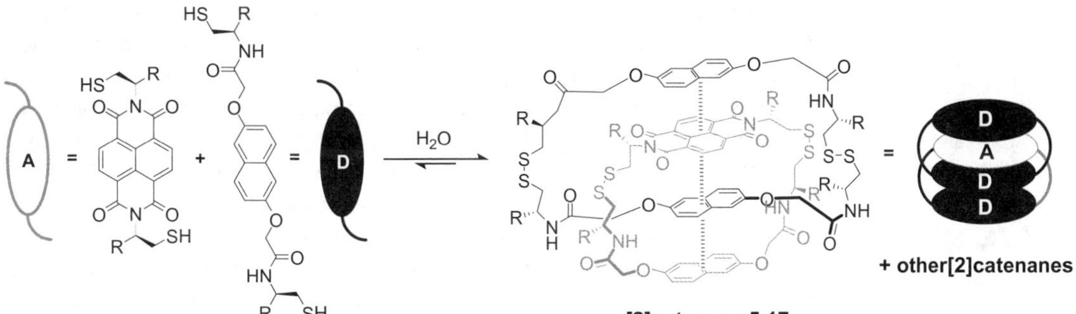

[2]catenane 5.17

Scheme 5.10 The self-assembly of π–π stacking MIMs under aqueous conditions causes deviation from the typical D-A-D-A [2] catenane arrangement due to the role of less discriminating hydrophobic interactions. D = π-donor, A = π-acceptor, R = CO_2H.

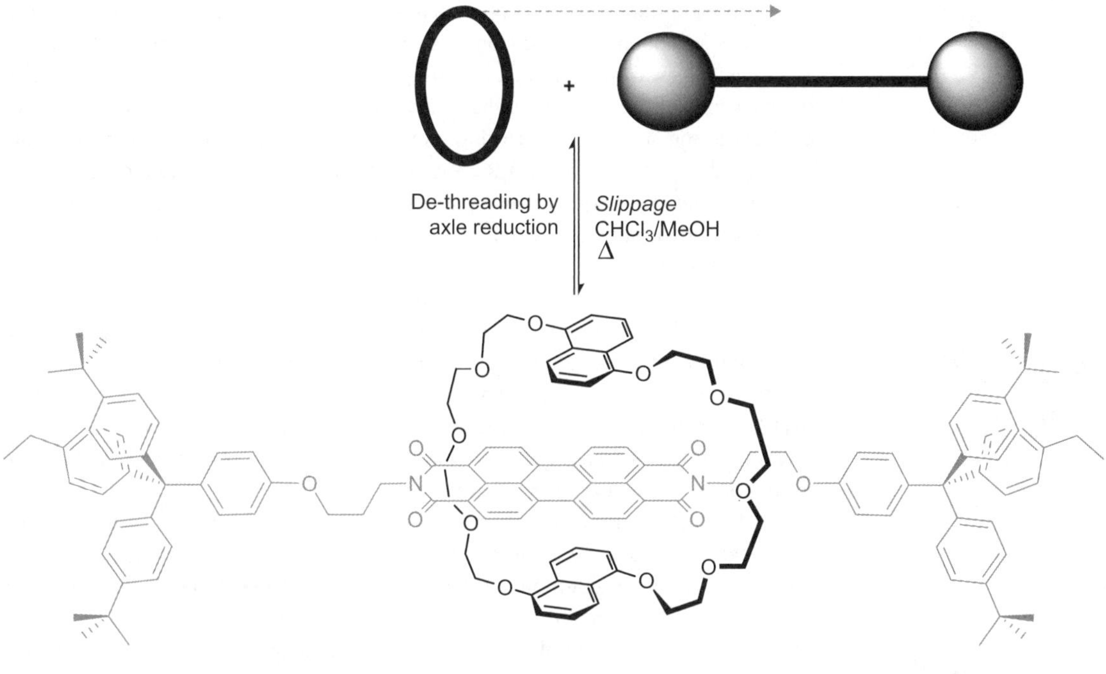

[2]rotaxane 5.18

Scheme 5.11 A [2]rotaxane prepared via a slippage strategy will be susceptible to de-threading at elevated temperatures and, in this case, electrochemical reduction.

Hydrogen bonding template-directed syntheses

Throughout this book we have witnessed the prominence of hydrogen bonding in supramolecular chemistry. MIMs are no different; they have also been prepared using hydrogen bonds as the primary templating interactions between molecular components. This work began when it was discovered that the condensation

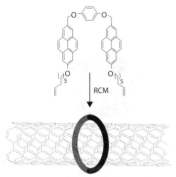

**Mechanically interlocked nanotube
(rotaxane 5.19)**

Scheme 5.12 The synthesis of a
rotaxane in which the axle component
is a carbon nanotube.

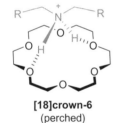

[18]crown-6
(perched)

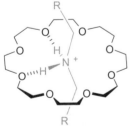

[24]crown-8
(interpenetrated)

Figure 5.9 The influence
of crown ether ring
size on formation of a
pseudorotaxane assembly
with a secondary alkyl
ammonium cation thread.

of a relatively rigid bis-amine with isophthaloyl chloride generated [2]catenane
5.20 in 34% yield when the macrocyclisation reaction was performed in a low
polarity solvent (Scheme 5.13). This occurs because the macrocycle precursors
become entwined by intermolecular hydrogen bonding during the condensation
reaction, templating interactions that 'live-on' in the resultant interlocked rings.
Therefore, mechanical bond formation is reliant on the correct orientation of the
hydrogen bond template interactions, an important condition also highlighted
by the synthesis of [2]rotaxanes containing benzylic amide-based macrocycles
(**5.21**, Scheme 5.13). Here, increasing the number of templating hydrogen bonds
or reducing the steric bulk increases the yield of the mechanically bonded product. Furthermore, switching out the glycylglycine units for a fumaramide motif
provides a near-quantitative route to [2]rotaxane **5.22** because ring–template
interactions benefit from preorganization of the trans-olefinic bond and strong
hydrogen bonding from the optimally positioned carbonyl groups, as evidenced
by IR spectroscopy.

Nicknamed the 'magic ring' approach, an unusual route to hydrogen bonded [2]
catenanes has been reported from a single macrocyclic starting material (Scheme
5.14). Moving away from amide condensation chemistry, the key here is the inclusion
of an alkene linkage in the benzylic amide macrocycle, enabling thermodynamically
controlled ring opening-ring closing metathesis in the presence of an appropriate
catalyst. Therefore, macrocycle **5.23** is converted into the corresponding homo[2]
catenane **5.24** in near quantitative yield. As expected, a low polarity solvent and high
reactant concentration are required so not to disrupt the templating intermolecular
hydrogen bonds.

Secondary dialkylammonium ions are amongst the most widespread recognition
units used in the synthesis of mechanical bonds. These organic cations are known to
form robust host–guest complexes with crown ether macrocycles via strong charge-
assisted $N^+H\cdots O$ hydrogen bonding interactions. While a secondary ammonium binds
to [18]crown-6 via a perched arrangement, interacting with only one face of the ring,
it was found that increasing the size of the crown ether to [24]crown-8 enabled the
formation of a pseudorotaxane (Figure 5.9).

Analysis of a variety of 1:1 macrocycle-ammonium thread pseudorotaxane
complexes indicate they are templated by multiple $N^+H\cdots O$ intermolecular hydrogen bonds, making assemblies strongly solvent and pH dependent. Non-com-
petitive solvents afford the strongest affinities, indeed dialkyammonium salts are
often solubilized in low polarity solvents upon their threading through the crown
ether's cavity, while weakly coordinating counteranions (e.g. PF_6^- or BF_4^-) limit the
competing effects of ion pairing. Due to greater host preorganization, association
constants for pseudorotaxane formation increase with decreasing crown ether ring
size, to a limit at which the crown ether backbone must distort to accommodate
the thread. While [24]crown-8 is considered to be optimal, the smallest [2]rotaxane
isolated to date (**5.25**) is comprised of a [20]crown-6-based macrocycle and an
ammonium ion axle, requiring only a phenyl ring and a CF_3 group to stopper it
(Scheme 5.15).

The relative simplicity, wide availability, and high yields of dialkylammonium templated mechanical bond formation highlight the potential for scale-up syntheses,

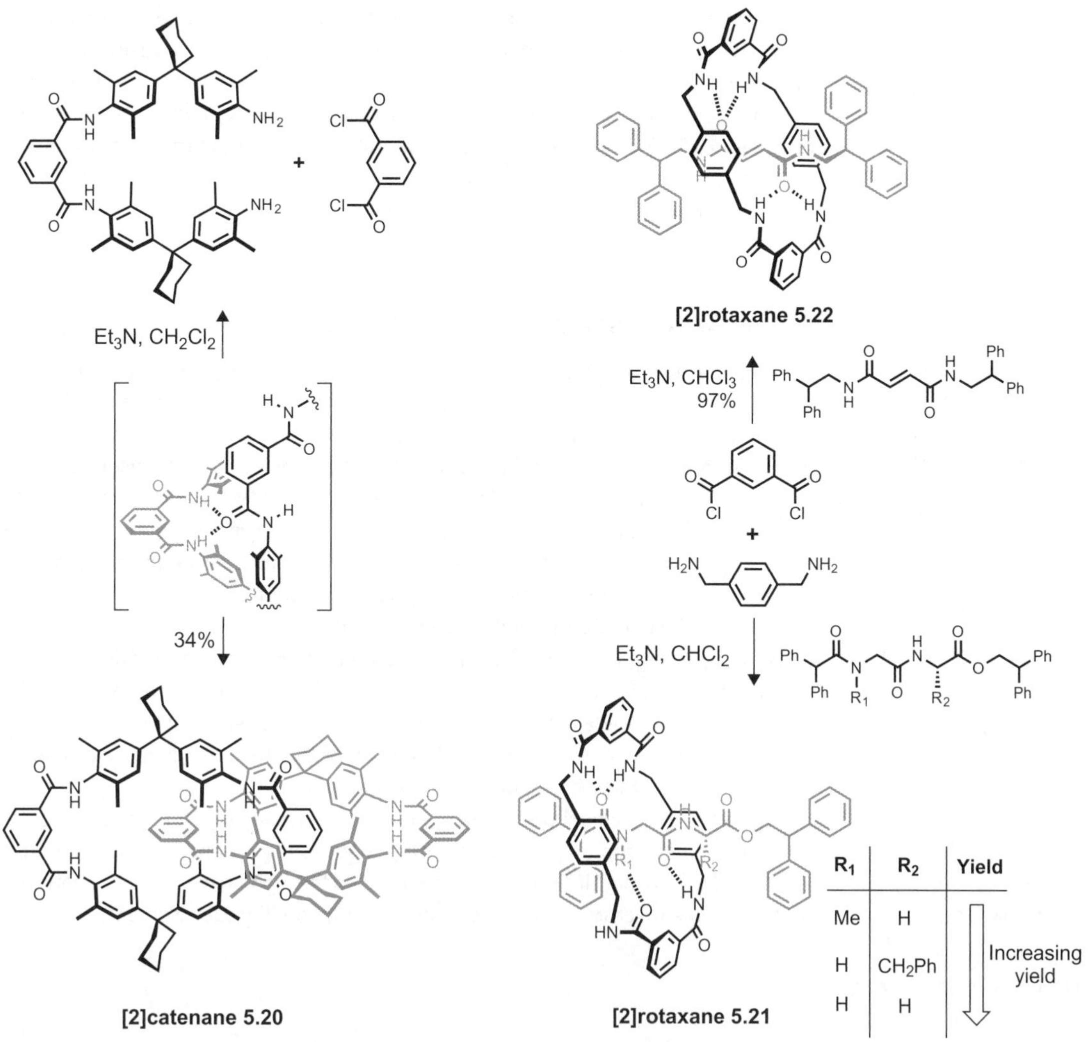

Scheme 5.13 Hydrogen bond interactions direct the synthesis of a neutral [2]catenane and [2]rotaxanes.

targeting potential use in industry. To this extent, MIMs have been prepared under solvent-free conditions in order to increase efficiency. Scheme 5.16 shows how the ball milling of a paracyclophane macrocycle, an alkylammonium-containing thread, and diaminonaphthalene stoppering units all together gave [2]rotaxane **5.26** in 49% yield.

In a similar vein, the threading and swelling/shrinking strategies of mechanical bond formation provide excellent atom economies and are typically reagent-less procedures. For example, a secondary ammonium cation-based thread containing two divinylcyclopropane termini is used to form a pseudorotaxane with dibenzo[24] crown-8, before gentle heating in nitromethane triggers a Cope rearrangement to generate larger cycloheptadiene stoppering groups and the [2]rotaxane **5.27** in 79%

5.23

'Magic ring'
synthesis
Grubbs' catalyst
>95%

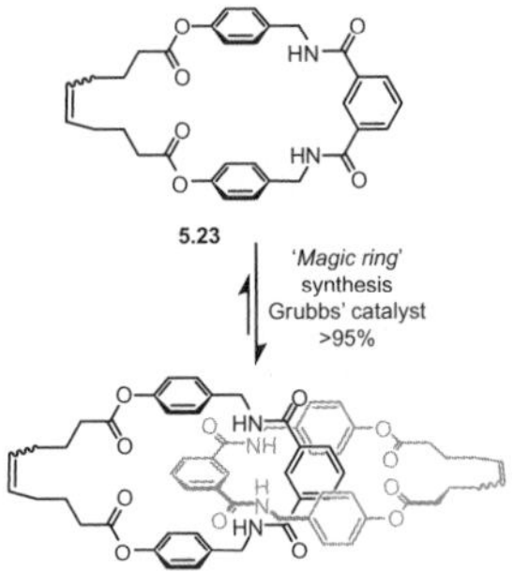

[2]catenane 5.24

Scheme 5.14 The 'magic ring' synthesis of a [2]catenane, afforded by Grubbs' catalysed ring opening-ring closing metathesis of a macrocyclic starting material and templated by intermolecular hydrogen bonding interactions.

B(Ar)$_4^-$

RCM,CH$_2$Cl$_2$ | Clipping
54%

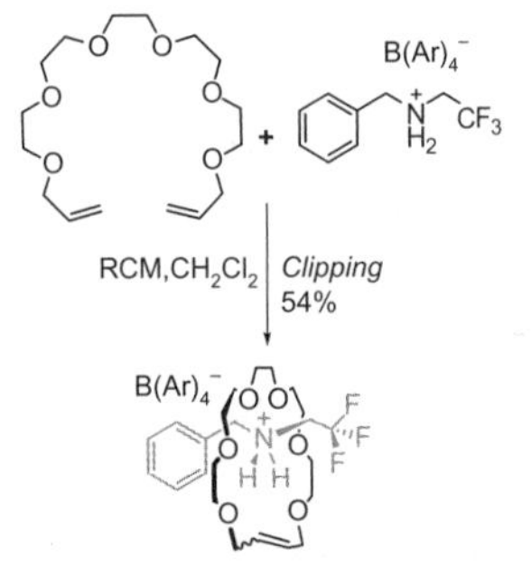

[2]rotaxane 5.25

Scheme 5.15 Secondary alkyl ammonium groups have been used to prepare one of the smallest known [2]rotaxanes.

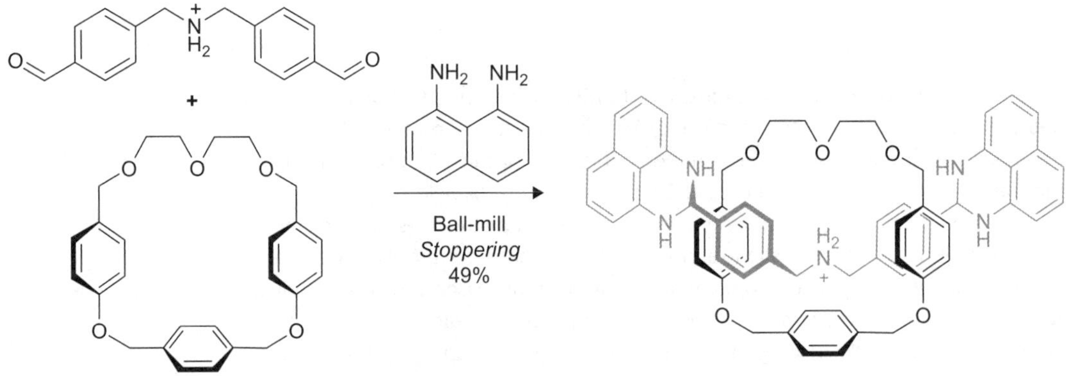

yield (Scheme 5.17). Alternatively, as outlined in Scheme 5.17, a pseudorotaxane is converted to a rotaxane by reducing the size of the macrocycle (i.e. shrinking). The photoextrusion of SO$_2$ using UV light is enough to kinetically trap a crown-ether based macrocycle on an dialkylammonium thread to form a hydrogen bond-templated [2] rotaxane **5.28**. This strategy requires a careful choice of stoppering groups; in this case, the cycloheptane groups are small enough to allow the large macrocycle to initially thread, but then large enough to prevent de-threading upon shrinking of the macrocycle by just a single atom.

Solvophobic template-directed syntheses

As apparent from the supramolecular cages seen in Chapter 4, solvophobic forces provide an effective tool to drive self-assembly and have been harnessed for the construction of mechanical bonds. Indeed, the hydrophobic-driven recognition of lipophilic guests by cyclodextrins (CDs) in aqueous media was well documented by the time it was initially explored for the construction of MIMs. Champions of 'green chemistry', CDs are naturally occurring macrocycles that were introduced in Chapter 3, Section 3.3. Moreover, their aqueous solubility, in particular permethylated CDs, make their compounds suitable for biological applications (Figure 3.16) and herein the opportunity to make mechanical bonds in water.

An early example saw α-CD forming a pseudorotaxane assembly with a mono-stoppered ferrocene alkyl ammonium thread in water before its conversion to [2] rotaxane **5.29** by amide coupling (Scheme 5.18). Interestingly, with ferrocene as one stoppering group and a naphthalene derivative as the other, the axle is asymmetric, meaning that the MIM is generated as two orientational isomers with respect to the direction the α-CD wheel points in. The non-directionality of hydrophobic interactions meant these isomeric interlocked compounds were generated in equal amounts, and two-dimensional NMR spectroscopy revealed through-space couplings between protons on the macrocycle and axle components to identify each isomer.

NH$_2$ NH$_2$

Ball-mill
Stoppering
49%

[2]rotaxane 5.26

Scheme 5.16 The solid state synthesis of a [2]rotaxane.

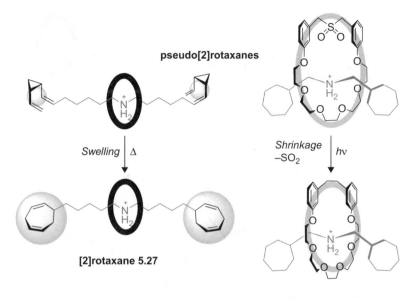

pseudo[2]rotaxanes

Swelling | Δ

Shrinkage | *hν*
−SO₂

[2]rotaxane 5.27

[2]rotaxane 5.28

Scheme 5.17 The preparation of [2]rotaxanes via swelling (left) and shrinking (right) strategies.

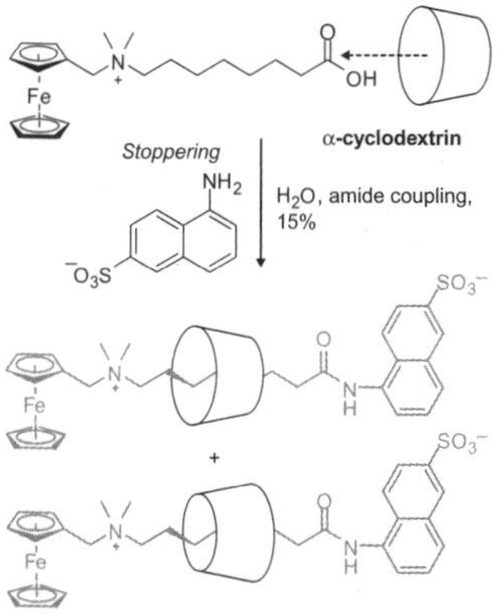

The strength of inclusion complexes formed between alkyl chains and CDs typically increases with chain length, making them potent solvophobic templates for forming pseudorotaxanes. This has been exploited for the synthesis of poly[*n*] rotaxanes, MIMs in which a single polymer chain is threaded through a series of macrocyclic rings generating multiple mechanical bonds (Scheme 5.19). To achieve this, an amine-terminated polyether is mixed with α-CD in water, triggering self-assembly into a polypseudorotaxane assembly which can be stoppered to give poly[*n*]rotaxane **5.30** in 60% yield. In addition to solvophobic forces, directional hydrogen bonding between complementary rims of CD macrocycles (i.e. 'head to head' and 'tail to tail') dictate the formation of a threaded channel-like structure.

Anion template-directed syntheses

Exploiting the wide variety of established anion recognition motifs employed in acyclic and macrocyclic hosts (Section 2.4), negatively charged species have since been employed as templates to direct the synthesis of interlocked architectures.

Anion templates for MIM syntheses were first developed by making use of the strong binding of a stopper phenolate derivative within the cavity of a tetralactam macrocycle via hydrogen bonding to an isophthalamide group. Subsequent S$_N$2 reaction with a bromo-benzylic functionalized electrophilic stopper delivered [2]rotaxane **5.31** in near quantitative yield (95%), due to persistent coordination of the phenolate anion within the macrocycle's cavity (Scheme 5.20).

Discrete anion templates have also been employed to promote the construction of interpenetrated architectures and synthesis of mechanical bonds. This concept

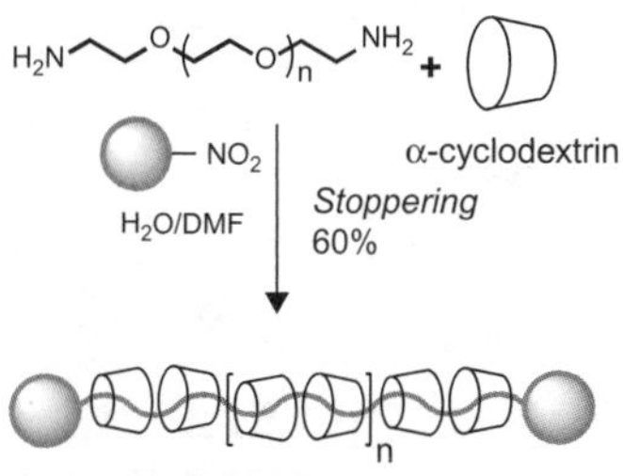

[2]rotaxane 5.29

Scheme 5.18 The hydrophobic effect is used to generate α-CD-based [2]rotaxanes with orientational isomers.

poly[*n*]rotaxane 5.30

Scheme 5.19 Hydrophobic interactions template the formation of poly[*n*]rotaxanes.

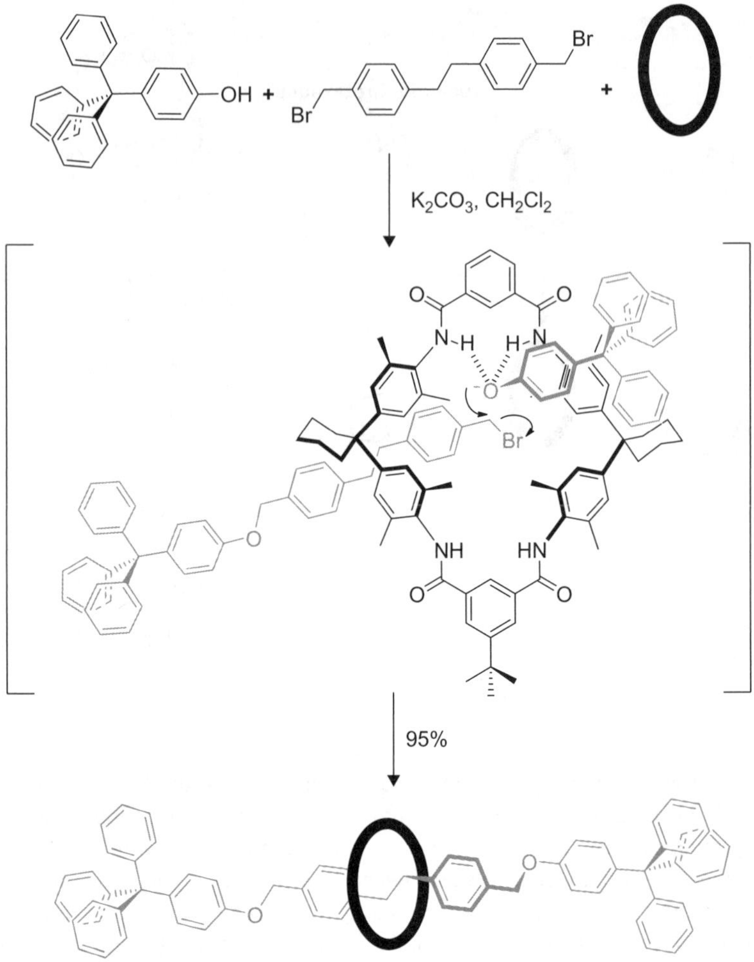

[2]rotaxane 5.31

Scheme 5.20 The synthesis of the first anion-templated [2]rotaxane used a snapping strategy.

originated from the realization that a tightly associated N-methyl pyridinium chloride ion-pair is coordinatively unsaturated, allowing an anion-binding macrocycle to coordinate the halide's vacant meridian, forming a pseudorotaxane assembly (Scheme 5.21). These assemblies are primarily stabilized by a network of hydrogen bonds and supported by aromatic stacking interactions. The association constants for these threaded assemblies are largest for chloride (K_a = 2,400 M^{-1} in acetone) over other anions due to charge density and size complementarity. This discovery facilitated the synthesis of the first chloride-templated [2]catenane (**5.32**) by a single ring-closing metathesis clipping reaction in 45% yield. As with Sauvage's Cu(I)-templated [2]catenane (Scheme 5.3), the chloride template can be removed, exposing a three-dimensional cavity that can be used to bind this halide with a high degree of selectivity. Further modifications to this approach have seen the use of sulfate, nitrate, and nitrite to template MIMs and the integration of redox- and photo-active reporter groups for anion sensing applications (see Figure 5.14).

The effectiveness of discrete anion templation to direct the formation of higher order interlocked architectures is demonstrated by the handcuff catenane in Scheme 5.22. This architecture comprises two covalently linked macrocycles that share a single, larger macrocycle that passes through both rings. The methodology chosen to synthesize **5.33** initially required formation of two chloride-templated pseudorotaxane assemblies (i.e. a pseudo[3]rotaxane), the threads of which are then connected up via ring-closing metathesis.

Outlined in Section 2.5, a growing number of artificial receptors are using halogen bonding as a non-covalent interaction for anion recognition. The contrasting properties of halogen bonding compared to hydrogen bonding (e.g. directionality and anion selectivity) provide new scopes for constructing MIMs. In particular, charge assisted CX···anion halogen bonding has proven effective in enabling halides other than chloride to be used as discrete anion templates. For example, the structure of [2]catenane **5.34** (Scheme 5.23) contains a bromide anion that is coordinated between two bromoimidazolium halogen bond donor groups on each of the macrocycle components. In the synthesis of **5.34**, it is these two linear halogen bonds that arrange the macrocyclic precursors around the halide template, enabling the MIM to be formed in 24% yield following a double-ring closing metathesis reaction. The importance of these halogen bonds is demonstrated by the fact that no interlocked product is produced if both counterions of the bromo-imidazolium-based ligands are non-coordinating (e.g. PF$_6^-$).

In contrast to the use of discrete anions, an organophosphate anion is integrated into the axle component of MIM **5.35** (Scheme 5.24). Stoppering of the alkyne-terminated negatively charged axle precursor using Cu(I)-catalysed alkyne–azide cycloaddition (CuAAC) 'click' chemistry in the presence of an excess of an anion binding

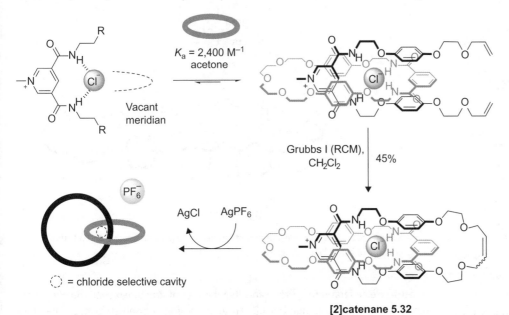

$K_a = 2,400$ M^{-1}
acetone

Vacant meridian

Grubbs I (RCM), CH$_2$Cl$_2$ 45%

PF$_6^-$

AgCl   AgPF$_6$

⬭ = chloride selective cavity

[2]catenane 5.32

Scheme 5.21 The synthesis of a [2]catenane by discrete anion templation. Removal of the template reveals a three-dimensional cavity for the selective recognition of chloride.

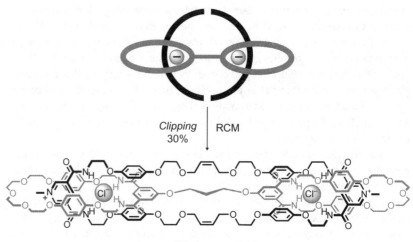

Clipping   RCM
30%

[2]catenane 5.33

Scheme 5.22 The anion-templated synthesis of a handcuff catenane.

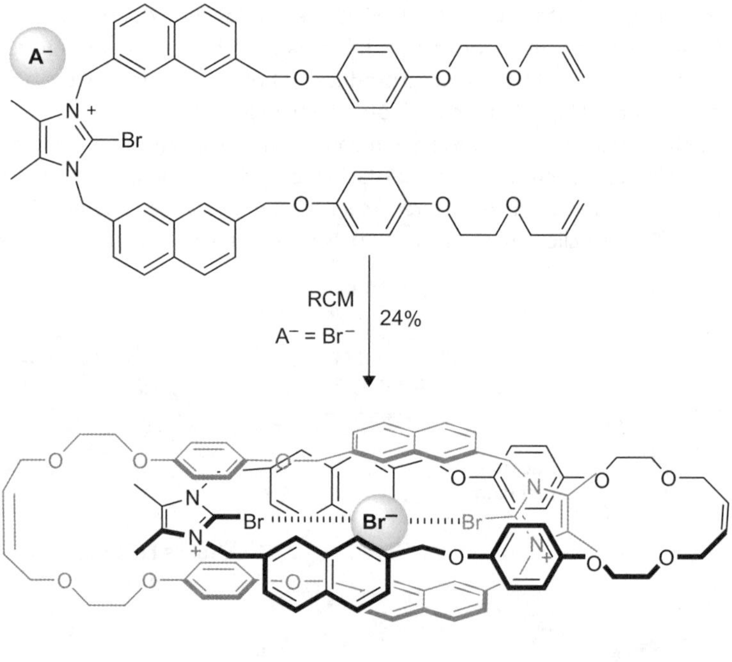

RCM   24%
$A^- = Br^-$

[2]catenane 5.34

Scheme 5.23 The bromide-templated synthesis of a [2]catenane that exploits the linearity of halogen bonding.

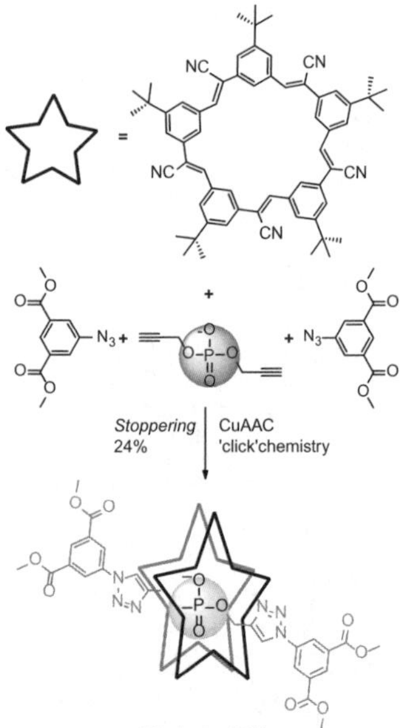

Stoppering │ CuAAC
24%   │ 'click'chemistry

[3]rotaxane 5.35

Scheme 5.24 The anion-templated synthesis of a [3]rotaxane, comprising two anion receptor macrocycles and a phosphate-based axle component.

macrocycle generates the [3]rotaxane. Its X-ray crystal structure reveals the phosphate anion supports a large number of CH··anion hydrogen bonds that must therefore guide formation of the 2:1 macrocycle:axle sandwich complex at the centre of the MIM. These interactions are also supported by aromatic stacking interactions between the macrocycle components.

Active metal template-directed syntheses

The primary role of the template presented so far in this chapter is to assemble and pre-organize the precursor reagents in a 'passive' fashion for mechanical bond formation. A more recent advance in the synthesis of mechanical bonds in high yields has employed a transition metal ion as both an organizing template and as a catalyst for the chemical reaction that covalently captures the molecular thread. This is called 'active metal tem-plation' and was first demonstrated using CuAAC 'click' chemistry, the same cycload-dition reaction seen in Scheme 5.24. This remains the most widely used reaction for active metal MIM syntheses today. In a typical mechanism, the Cu(I) metal ion is bound within the macrocycle's cavity, in its preferred tetrahedral coordination geometry, al-lowing it to first template formation of the interpenetrated assembly and then catalyse the formation of the covalent bond between the azide- and alkyne-functionalized axle

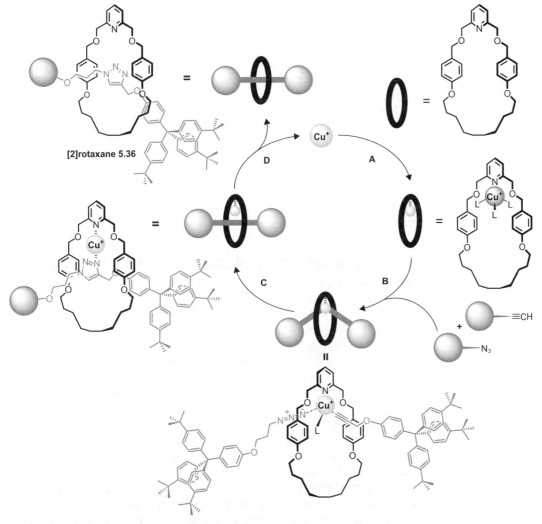

Scheme 5.25 The general mechanism of active metal templation, demonstrated here in the preparation of a [2]rotaxane using CuAAC chemistry; A) the Cu(I) catalyst is initially bound within the cavity of the macrocycle; B) coordination of the azide and alkyne axle-forming reagents (L = catalyst ligand); C) endotopic reaction to generate the mechanical bond; D) the catalyst is re-generated and metal-free [2]rotaxane produced.

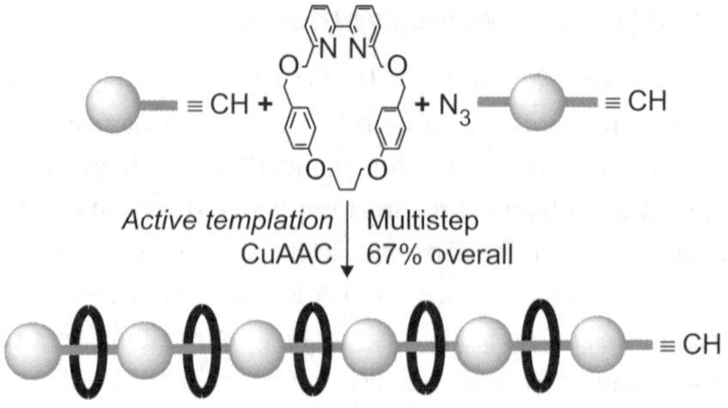

[6]rotaxane 5.37

Scheme 5.26 Active metal templation is key in the synthesis of a [6]rotaxane in high yield.

precursors, capturing the interlocked structure (**5.36,** Scheme 5.25). The propensity for chelation of the macrocycle and axle precursors to the Cu(I) catalytic centre in an endotopic coordination mode (i.e. within the macrocycle's cavity) makes the reaction amenable to forming a mechanical bond and is critical to the high yields of rotaxanes and catenanes produced by this synthetic strategy.

A subsequent investigation into the effect of ring size on CuAAC active metal-templated MIM syntheses revealed smaller Cu(I)-binding macrocycles containing 2,2′-bipyridyl groups can increase yields, ultimately delivering a [6]rotaxane (**5.37**) in 67% overall yield (Scheme 5.26). In addition to Cu(I)-based syntheses (e.g. CuAAC, Eglington–Glaser–Hay), other transition metal-catalysed reactions, including Pd(II)-catalysed oxidative Heck coupling, Zn(II)-promoted Diels–Alder reaction and Ni(II)-mediated homocoupling, have been exploited for making mechanical bonds by active metal templation.

5.3 Properties and applications of mechanically interlocked molecules

Establishing effective synthetic routes to rotaxanes and catenanes has enabled chemists to probe the structure–function relationship of mechanical bonds in more detail, uncovering unique properties that can be put to use in a variety of applications.

The chemical and physical properties of MIMs are often different to those of their non-interlocked components. The influence of the mechanical bond is particularly evident in the nature of non-covalent interactions between MIM sub-components; where close proximity, yet relative flexibility, means these interactions may be considered as hybrids between those found in purely intermolecular (e.g. a host–guest complex) or intramolecular (e.g. an organic molecule) systems. As such, this section begins by examining and highlighting the distinct supramolecular chemistry of MIMs. This is followed by a discussion on the influence of mechanical bonds on photophysical and electrochemical properties. Mechanical bonding can influence the sterics and electronics of functional groups present in the molecular sub-components, providing the potential to modify intrinsic chemical reactivity without altering covalent structure. As such, a subsection has been devoted to the successful applications of MIMs as catalysts and reagents, including those MIMs with optical activity. The synthesis and properties of MIM materials are also discussed.

Supramolecular properties of mechanically interlocked molecules

While mechanically interlocked molecules are not considered true supramolecular assemblies like a host–guest complex, they will often possess non-covalent interactions between their molecular sub-components. The following examples highlight how the properties of the mechanical bond can influence the supramolecular chemistry of MIMs, enabling them to stabilize reactive species, alter solubilities or raise-the-bar of molecular recognition.

It was described in Section 5.2 how Sauvage, on synthesizing the Cu(I)-containing [2]catenane **5.4**, noted the significant stabilization of MIMs to demetalation in comparison to non-interlocked metal–ligand complexes (Scheme 5.3). The interlocked architecture acts to preorganize the multidentate ligands of these rotaxanes, providing mechanical chelation, or a 'catenand effect'. Furthermore, the mechanically bonded ligand set was found to disfavour oxidation of the Cu(I) template because the large degree of steric crowding around the metal centre restricts geometrical rearrangements that would otherwise stabilize the higher oxidation state.

More recently, the coordination of a range of transition metal ions to the [2]rotaxane-based ligand set **5.38** has been investigated (Scheme 5.27). The steric constraints of the interlocked cavity results in unusual coordination environments being imposed on transition metal complexes. For example, in the Cu(II) complex of **5.38**, the metal ion possesses a highly distorted tetrahedral geometry, which is rare and closely resembles copper ions found in the crowded active site of a protein rather than the typical Jahn–Teller-dictated tetragonal geometry. Furthermore, the rigidity of the interlocked ligands was shown to prevent ligand reorganization during electrochemical oxidation or reduction; behaviour familiar from metallo-enzymes in which redox-induced structural changes are reduced to enable efficient intermolecular electron transfer processes.

Catenane **5.4** and rotaxane **5.38** were shown to enhance the stability of transition metal ions in low oxidation states. This can be considered a more general property of MIMs, namely their ability to stabilize reactive species and influence chemical reactivity, via manipulation of non-covalent interactions within the interlocked framework. Indeed [2]rotaxane **5.39** was used to stabilize a compound that is otherwise undergoes thermal decomposition (Scheme 5.28). Fullerenes substituted with pyrrolidine N-oxide groups are normally only stable over the course of several days at –20 °C. However, the encapsulation of this axle stoppering species by a hydrogen bond-donating interlocked macrocycle in rotaxane **5.39**, inhibits de-oxygenation, enabling these materials to be exploited for organic solar cells with enhanced lifetimes and efficiencies. In the future, mechanical bonds might also be harnessed for capturing reactive intermediates in order to uncover new reaction mechanisms.

A common outcome of mechanical bonding is an increase in the strength of non-covalent interactions between molecular components upon their interlocking, particularly within small, preorganized MIMs. For example, strong intramolecular hydrogen bonding is responsible for lowering the acidity of the secondary dialkylammonium group present in the axle component of crown ether-based rotaxanes such as **5.40**. As such, their attempts to neutralize **5.40** using bases known to deprotonate the analogous pseudorotaxane were unsuccessful (Scheme 5.29). This effect could be quantified by dissolving **5.40** in D_2O and monitoring the rate of H/D exchange by NMR spectroscopy.

The saturation of non-covalent interactions can drastically alter the solubilities of MIM components. For example, the solubility of an axle can be increased via its encapsulation by a macrocycle in a rotaxane. Poly(ethylenedioxythiopehene)

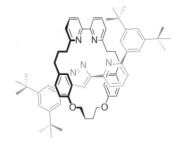

[2]rotaxane 5.38

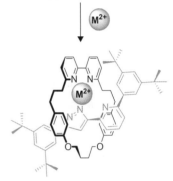

M = Cu, Ni, Co, Zn

Scheme 5.27 MIMs have been engineered with sterically crowded and rigid ligand frameworks to promote unusual geometries of transition metal cations.

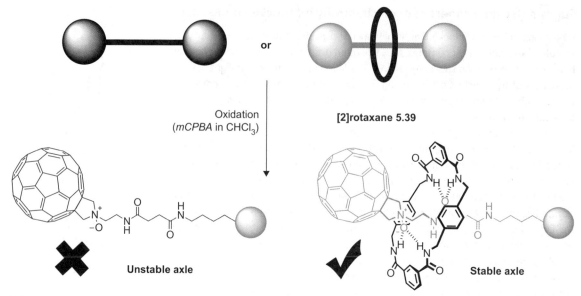

or

[2]rotaxane 5.39

Oxidation
(*mCPBA* in CHCl₃)

Unstable axle

Stable axle

Scheme 5.28 A reactive fullerene pyrrolidine *N*-oxide is stabilized within a [2]rotaxane architecture via strong intramolecular hydrogen bonding.

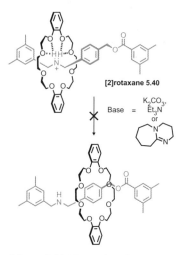

[2]rotaxane 5.40

Base = K₂CO₃,
Et₃N
or

Scheme 5.29 The acidity of a secondary ammonium group is lowered upon its integration into a [2] rotaxane architecture due to strong inter-component hydrogen bonding.

5.41 is an electrically conducting polymer that, due to its extensive conjugated aromatic structure, is susceptible to strong intermolecular aromatic stacking interactions, making it poorly soluble in most solvents. However, by use of its strong solvophobic interaction with cucurbit[7]uril macrocycles, the transformation of this polymer into poly[*n*]rotaxane **5.42** significantly enhances aqueous solubility because the polymeric backbone is shielded from the solvent (Scheme 5.30). Polyrotaxane **5.42** is hence preferable to polymer **5.41** for the fabrication of electronic devices because it enables less energy-intensive and more environmentally friendly handling of the material in water.

Earlier examples illustrated how mechanically bonded ligand sets can strengthen the coordination of transition metal cations (Scheme 5.27). As a natural extension to this, MIMs have also been engineered for strong and selective binding of negatively charged and neutral guests. Critically, the constrained three-dimensional cavities formed between their interlocked components, bearing resemblance to the preorganized binding pocket of protein active sites, prime these synthetic hosts for molecular recognition. We have already seen an example of this in the the isophthalamide-based [2]catenane **5.32** that exhibited preferential binding of chloride (Scheme 5.21). Likewise, the larger anion binding motif bis(triazolium)acridine incorporated into [2]rotaxane **5.43** delivers enhanced recognition of nitrate (Scheme 5.31). Idealized size and shape complementarity, optimized for charge assisted hydrogen bonding between the interlocked host and oxoanion guest, are responsible for strong binding in a competitive, mixed organic-aqueous, solvent medium (K_a = 1,000 M^{-1}). The nitrate anion represents an important target for extraction from natural aquatic environments; the leaching of nitrate-containing fertilizer pollutants causes eutrophication and subsequent depletion of oxygen in the water which is detrimental to aquatic life cycles.

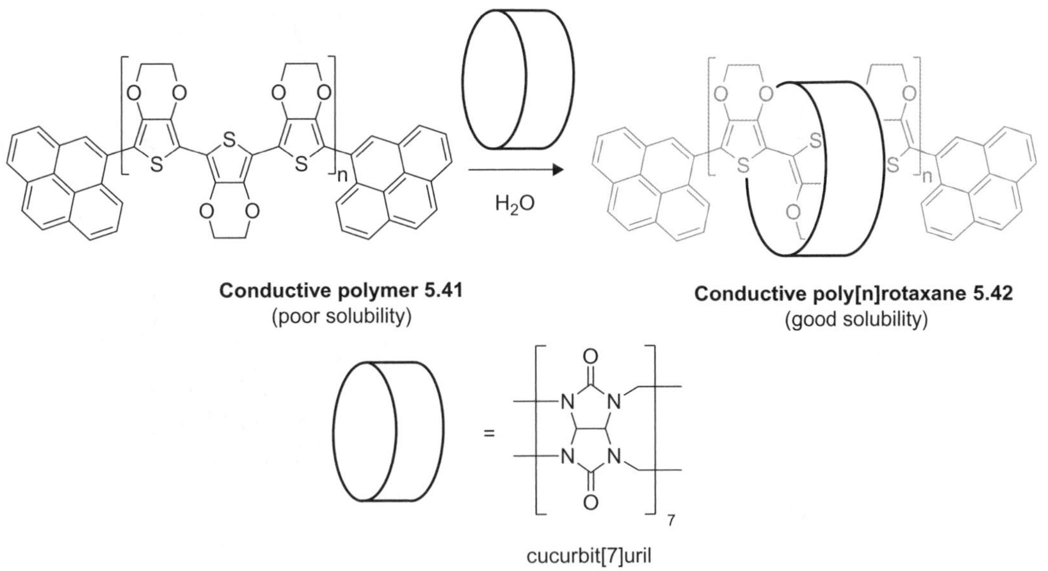

Conductive polymer 5.41
(poor solubility)

Conductive poly[n]rotaxane 5.42
(good solubility)

cucurbit[7]uril

Scheme 5.30 An electrically conducting polymer can be solubilized in water upon its encapsulation by hydrophilic macrocycles in a poly[n]rotaxane architecture.

It is also possible to fashion a linear binding cavity from a rotaxane framework. The [2]rotaxane **5.44** is a receptor with a cylindrical cavity and geometric match for cyanate, dictating selectivity for this anion over spherical- and trigonal planar-shaped guests (Scheme 5.31). Interestingly, **5.44** can also act as a heteroditopic host, binding tetra-alkyl ammonium cations within the calix[4]pyrrole unit, making it applicable for the extraction of salts (i.e. ion-pair binding).

The recognition of naturally occurring chiral substrates such as amino acids is also important. In comparison to acyclic receptors, MIMs provide a unique opportunity to construct a crowded, yet flexible, chiral cavity to enhance enantioselectivity of binding. Primarily this has been realized via the incorporation of stereogenic centres or axially chiral groups into one or more of the interlocked components within the binding cleft. For example, homo[2]catenane **5.45** comprises two 1,1'-binaphthyl-2,2'-diyl hydrogenphosphate-based macrocycles to create a cavity with S,S-axial chirality for the enantioselective recognition of chiral diamine guest molecules (Figure 5.10). Stereoselectivity for the D-isomer of arginine was observed via hydrogen bonding interactions from the chiral hydrogenphosphate groups of **5.45** that orientate the chiral guest to favour the binding of one enantiomer over the other. Further work into rotaxane hosts exhibiting stereoisomerism has indicated mechanically bonded systems outperform their non-interlocked constituent parts in terms of enantioselectivity for amino acids, whilst the inclusion of a chiral group within the macrocycle component appears to be more important than the axle.

Photo- and redox-active mechanically interlocked molecules

The restricted environment of the mechanical bond can alter the photo- and redox-properties of functional groups appended to the components of interlocked molecules.

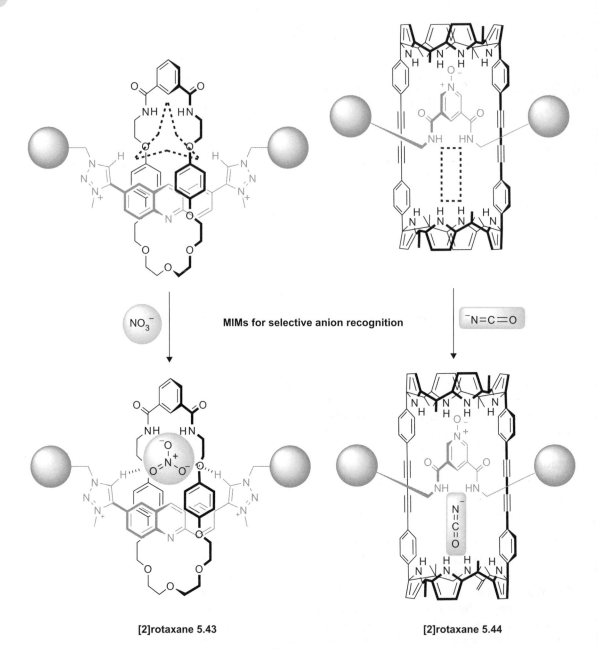

MIMs for selective anion recognition

[2]rotaxane 5.43

[2]rotaxane 5.44

Scheme 5.31 Selective anion recognition is afforded by [2]rotaxane hosts courtesy of their preorganized interlocked cavities with well-defined complementary shapes for guests.

This has been exploited to great success in the development of robust optical imaging agents and sensors.

Squaraines are molecular dyes that absorb in the near-infrared region of the electromagnetic spectrum, making them highly applicable for *in vivo* biomedical imaging because low energy radiation is less dangerous and has greater tissue penetration. There-

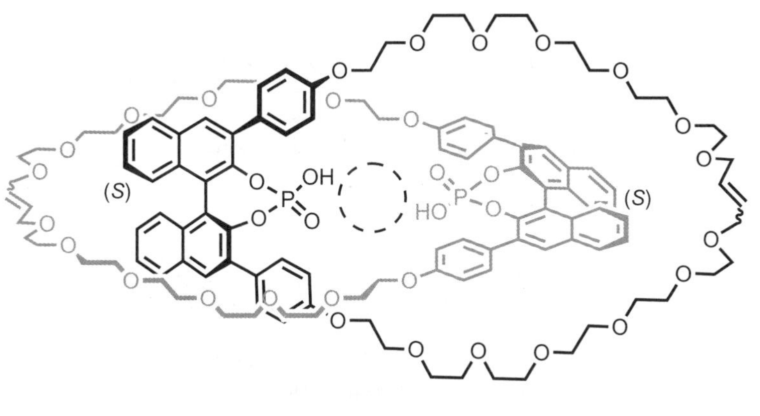

Chiral [2]catenane 5.45

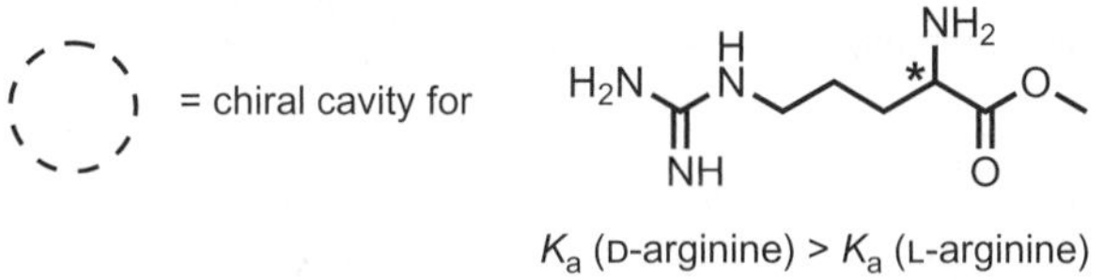

= chiral cavity for

K_a (D-arginine) > K_a (L-arginine)

Figure 5.10 The incorporation of 1,1'-binaphthyl-2,2'-diyl hydrogenphosphate groups with axial chirality into the macrocycle components of a [2]catenane enables the enantioselective recognition of the amino acid D-arginine.

fore, in an effort to enhance the stability and suitability of squaraines, this dye has been incorporated into the axle component of MIMs such as [2]rotaxane **5.46**, where strong intramolecular hydrogen bonding locks the macrocycle component over the chromophore providing it with a protective shield (Figure 5.11). Macrocycle encapsulation inhibits aggregation-induced broadening of the squaraine absorption spectrum and reduces the risk of degradation by nucleophilic species. By contrast, an unshielded axle component **5.47** is highly susceptible to both these processes under biological conditions.

Cyanine is a common fluorescent probe, but has challenges associated with its longevity in biological environments due to a propensity to react with singlet oxygen upon excitation, a detrimental and irreversible effect known as photobleaching. To address this, a cyanine dye-containing axle has been mechanically interlocked within the cavity of an α-cyclodextrin macrocycle, delivering a 40-fold increase in photostability relative to the free fluorophore in water (**5.48**, Figure 5.12). This effect is concomitant with an increase in the fluorescence quantum yield of the [2]rotaxane; rationalized by the reduced flexibility of the encapsulated dye. Poly[n]rotaxane analogues of this MIM have also been prepared in which an emissive conjugated polymer, poly(para-phenylene), is threaded through multiple cyclodextrin units (Figure 5.12). These compounds are termed 'insulated molecular wires' because the polymer's electro- and photo-chemical properties are maintained despite inclusion within a sugar-based 'sheath'. By inhibiting aggregation, the poly[n]rotaxane **5.49** delivers an emission enhancement and a spectral shift to shorter wavelengths, hence enabling their use as materials for blue organic light emitting diodes (OLEDs).

A different type of MIM has been developed in which photo-active motifs are incorporated into both the axle and macrocycle components of a [2]rotaxane, making it capable of exciplex emission (**5.50**, Scheme 5.32). An exciplex is a short-lived

Squaraine axle 5.47

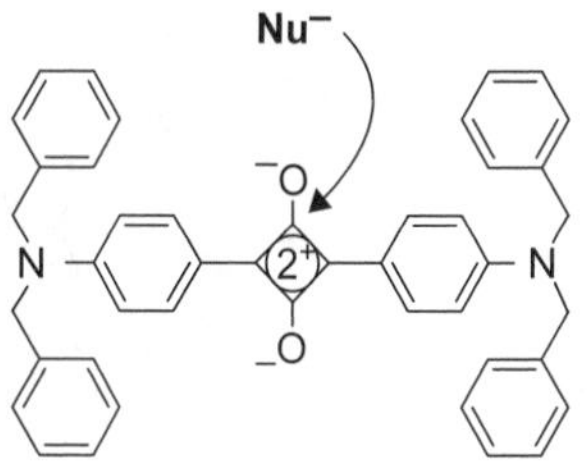

Dye is unshielded

vs.

Squaraine [2]rotaxane 5.46

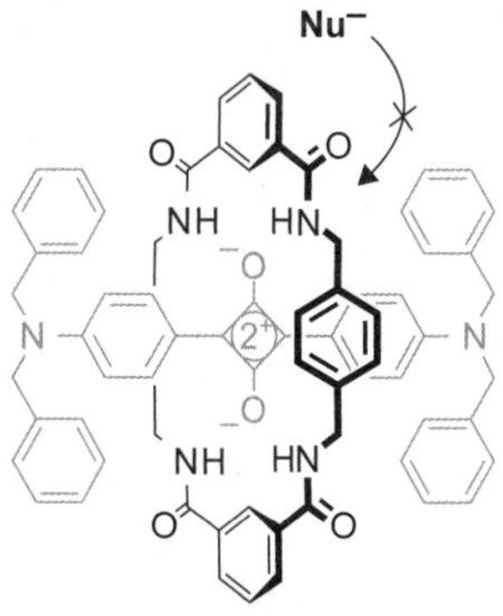

Dye is shielded by macrocycle

Figure 5.11 The encapsulation of a squaraine dye within a [2] rotaxane makes it more applicable for bioimaging because the macrocycle shields the dye from nucleophilic (Nu⁻) attack.

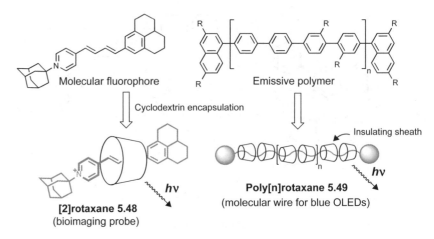

Figure 5.12 The incorporation of organic fluorophores into cyclodextrin-based rotaxane frameworks enhances photophysical properties for various applications.

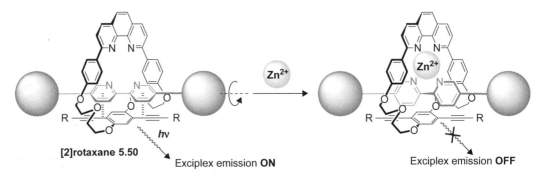

Scheme 5.32 A photo-active [2]rotaxane produces exciplex emission that can be used for the sensing of Zn(II).

non-covalent complex formed between two different molecular components in the excited state but not in the ground state. Photoexcitation of [2]rotaxane **5.50** generates red-shifted exciplex emission due to π–π electronic communication between the co-facial aromatic groups in macrocycle and axle components. Interestingly, the nature of the mechanical bond is responsible for the unique photophysical properties of this system; specifically the traits of both an intermolecular (i.e. spatial separation) and intramolecular (i.e. no concentration dependence) exciplex assembly. The MIM **5.50** can also be used as an optical sensor for Zn(II) since binding of the metal cation prevents exciplex formation and quenches the emission (Scheme 5.32). Following this rationale, a variety of photophysical MIM sensors have been developed that, upon selective substrate recognition, illicit a macroscopic response courtesy of an optical reporter group integrated within, or proximal to, the binding pocket.

Redox-active acyclic and macrocyclic receptors have been used for the electrochemical sensing of charged guest species and this has also been extended to MIMs via the integration of redox-active groups. Notably, Os(II)–bipyridyl provides both an electrochemical and luminescent reporter group and its use in a [2]rotaxane has enabled multimodal anion sensing to be performed via communication between interlocked cavity binding site and reporter group. Furthermore, the interfacing of chemical sensors with solid electroactive (e.g. gold) or optically transparent (e.g. glass) surfaces lies central to the construction of sensory devices. Towards this goal, Os(II)–bipyridyl [2]rotaxane **5.51** was immobilized onto a gold electrode surface whereby chloride recruitment is detectable via a cathodic shift in the Os(II/III) redox couple (Figure 5.13).

Mechanically interlocked molecules as catalysts and reagents

Mechanically interlocked molecules have been applied as novel reagents and catalysts in chemical transformations, displaying distinct reactivity in comparison to their separate non-interlocked sub-components.

The variable oxidation states and coordination numbers of transition metals has made them ubiquitous in catalysis. This has been exploited by the Cu(I)-binding [2]rotaxane **5.52** in which the mechanically bonded framework promotes an intramolecular cycloaddition reaction (Scheme 5.33). The diyne axle undergoes a rapid cycloaddition reaction with aniline in the presence of a sub-stoichiometric amount of Cu(I), generating a pyrrole-based rotaxane analogue. Despite the greater steric bulk, this reaction was significantly faster with the MIM reagent than axle **5.53** alone because the rotaxane coordinates the Cu(I) catalyst endotopically, positioning it proximal to the axle and so enhancing reactivity.

The coordination of a transition metal ion at a site peripheral to the interlocked cavity has enabled the mechanical bond to provide contrasting catalytic reactivity compared to non-interlocked analogues. A metal-based [2]rotaxane catalyst (**5.54**) has been developed which can be switched on using a chemical cofactor, a process known as allosteric control that is common in enzymes. Containing

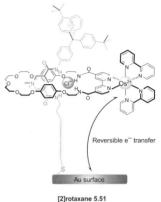

Reversible e⁻ transfer

[2]rotaxane 5.51

Figure 5.13 An Os(II)–bipyridyl-based [2]rotaxane is fabricated on a gold surface where it acts as an electrochemical sensor for chloride.

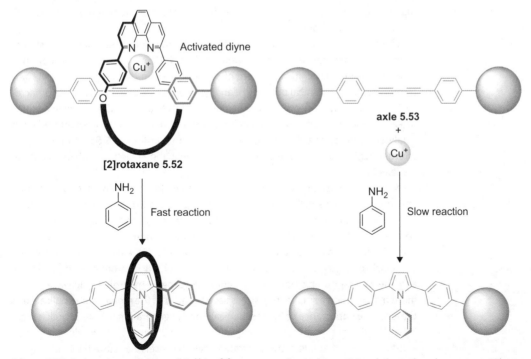

Scheme 5.33 Endotopic coordination of Cu(I) in a [2]rotaxane accelerates its reactivity relative to the axle component alone.

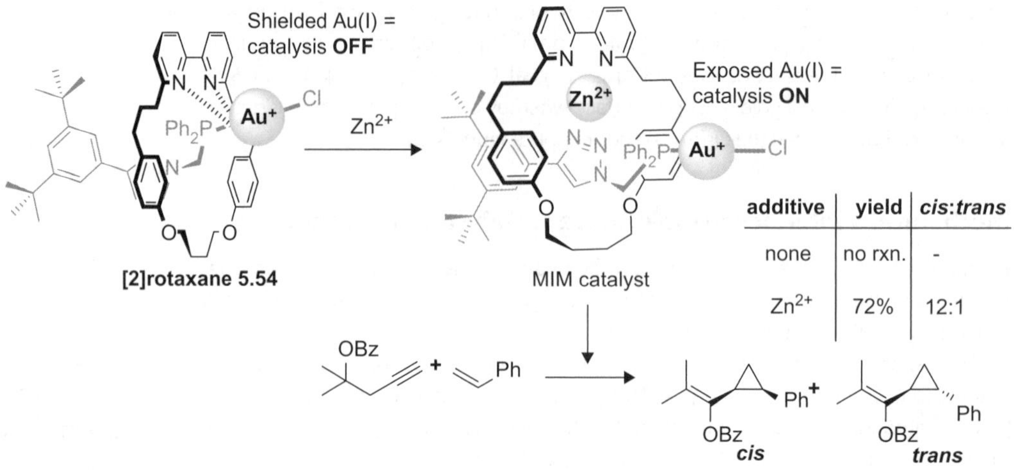

Scheme 5.34 A Au(I)-based [2]rotaxane acts as a catalyst, the activity of which can be controlled through coordination of a Zn(II). Figure adapted under the terms of the Creative Commons Attribution License from M. Galli et. al., *Angew. Chem. Int. Ed.*, 2015, **54**, 13545.

additive	yield	cis:trans
none	no rxn.	-
Zn²⁺	72%	12:1

an Au(I) complex as one of its stoppering units, [2]rotaxane **5.54** was initially shown to be inactive for catalysing the cyclopropanation of styrene due to the metal ion being coordinatively saturated and sterically shielded via its interaction with the bipyridyl-based macrocycle (Scheme 5.34). However, the addition of Zn(II) ions displaces Au(I) from the macrocycle cavity, switching on Au(I) catalytic activity. The [2]rotaxane delivers higher yields and greater diastereoselectivity in the cyclopropanation reaction than the axle alone, which is rationalized by the enhanced steric hindrance provided by the mechanical bond.

The source of chirality in the majority of rotaxanes and catenanes is the stereochemistry of their covalent sub-components (e.g. **5.45** in Figure 5.10). Less common is mechanical planar chirality, a form of stereoisomerism that arises in a [2]rotaxane formed by the mechanical bonding of an asymmetric axle and an asymmetric macrocycle (Figure 5.14). Harnessing this, the previous MIM catalyst **5.54** has been adapted to exhibit mechanically planar chirality (**5.55**) and hence used for enantioselective catalysis. Outlined in Figure 5.14, the enantiopure [2]rotaxane **5.55** (*R*-isomer) now affords enantioselective catalysis of the cyclopropanation reaction, with excellent stereoselectivities.

Negating the use of transition metals altogether, the steric constraint of a secondary amine within a small rotaxane framework has been exploited for the heterolytic activation of dihydrogen. In [2]rotaxane **5.56**, the macrocyclic ring surrounding an aniline Lewis base prevents adduct formation with a boron Lewis acid, generating a species known as a frustrated Lewis pair (Scheme 5.35). As a consequence of this unquenched reactivity, the exposure of **5.56** to H₂ produces a protonated [2]rotaxane and the corresponding borohydride anion. Hydrogen heterolytic cleavage occurs more readily when a smaller ring size is used because of the greater steric bulk around the axle. The importance of the mechanical bond in generating this steric bulk is highlighted by the fact small molecule activation was not observed by the axle component alone. In terms of applications, frustrated Lewis pairs serve as sustainable metal-free

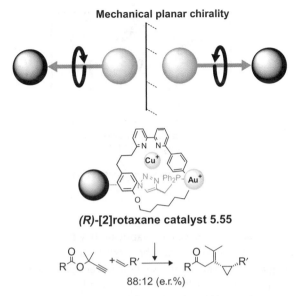

Mechanical planar chirality

(R)-[2]rotaxane catalyst 5.55

88:12 (e.r.%)

Figure 5.14 Akin to the configurations formed by covalent molecules with planar chirality, [2]rotaxanes with asymmetric axle and macrocycle components have two possible enantiomers. This mechanical planar chirality has been exploited for enantioselective catalysis.

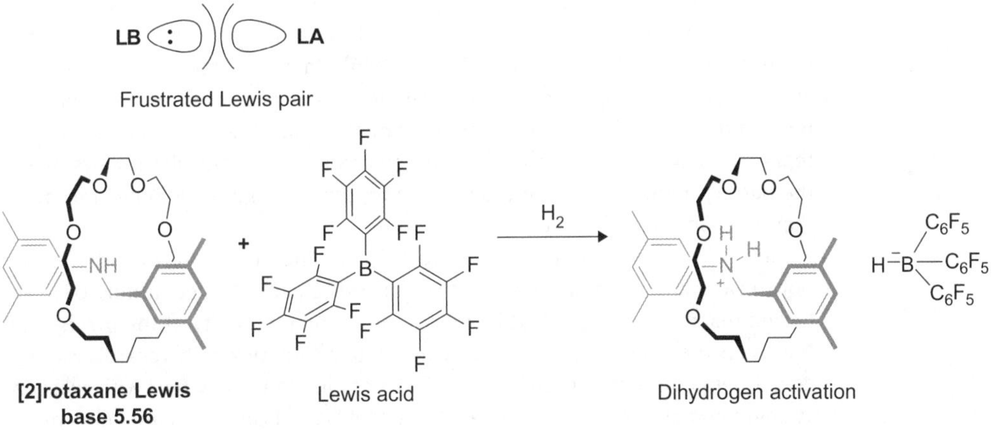

LB ⌣ : ⌣)(⌢ ⌢ LA

Frustrated Lewis pair

[2]rotaxane Lewis base 5.56   Lewis acid   Dihydrogen activation

Scheme 5.35 A [2]rotaxane is exploited as the Lewis base component of a frustrated Lewis pair for H_2 activation.

catalysts for synthetic hydrogenations while reversing H^+/H^- separation presents new possibilities in hydrogen storage.

Mechanically interlocked molecules in materials

In addition to solution phase chemistry, MIMs have also found applications as materials with properties contrasting their non-interlocked analogues.

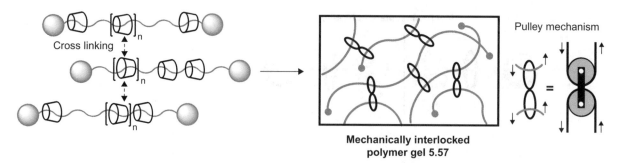

Pulley mechanism

Mechanically interlocked
polymer gel 5.57

Figure 5.15 The formation of a mechanically interlocked gel by cross-linking poly[*n*]rotaxanes generates a highly flexible material due to the pulley-action of the mobile polymer chains.

Poly[*n*]rotaxane-based gels have been fabricated by threading α-cyclodextrins onto polyethylene glycol axles and then cross-linking via the macrocycle components (Figure 5.15). This presents an unconventional type of cross-linking because the polymer chains (axles) are not covalently connected nor are there strong intermolecular interactions between them as seen previously in supramolecular gels (Section 4.3). Instead, unique 'figure of eight' cross links are produced that can pass freely along polymer strands allowing them to equalize chain tension in a cooperative fashion, much like the mechanism of a pulley. As a result, the gel is flexible and tensile to twice its length which, in addition to its transparency, makes **5.57** appropriate as an anti-scratch coating for displays. These physical properties are a direct consequence of the mechanical bond 'pulley effect' because cross-linking a non-interlocked mixture of the same molecular components does not lead to gelation. Related poly[*n*]rotaxane materials are also finding application as binding agents in the expanding silicon anodes of rechargeable batteries. While the effectiveness of conventional binders gradually decreases upon repeated charging and discharging, the mechanism of molecular pulleys makes the battery materials much more resilient to large changes in volume, improving their efficiency and lifetime.

Related to poly[*n*]rotaxanes, Figure 5.16 outlines the synthesis of the first poly[*n*]catenane material in which the polymer chains consist entirely of interlocked macrocyclic rings (**5.58**). The conformational mobility of the mechanical bonds ensures these topological polymers retain their flexibility regardless of the constituent covalent ring, giving a new class of functional polymers with excellent strength and elasticity. These materials are highly applicable as dampening materials or elastomers.

Finally, MIMs have been used as building blocks in metal–organic framework materials (see Section 4.2) by incorporating carboxylate-stopped [2]rotaxanes such as **5.59** as the organic linker between transition metal ion nodes (Figure 5.17). Interestingly, the inherent dynamic properties of the rotaxane remain intact because the macrocycle component has enough space within the rigid framework to be able to move around or along its axle component. This was characterized by solid state NMR spectroscopy. The molecular motion can be switched off if the void space is occupied by solvent guest molecules. The ability to arrange MIMs in a dense and ordered array, whilst also controlling the dynamics of their interlocked components, opens

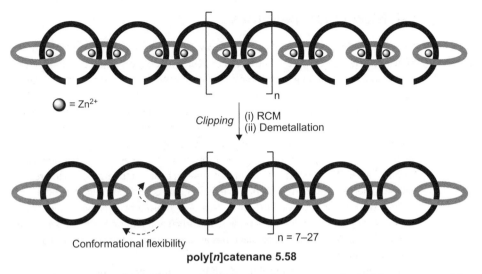

Figure 5.16 A metal template-directed synthesis is used in the preparation of a poly[*n*]catenane, materials with high elasticity due to the dynamic properties of the mechanical bonds linking the monomer units.

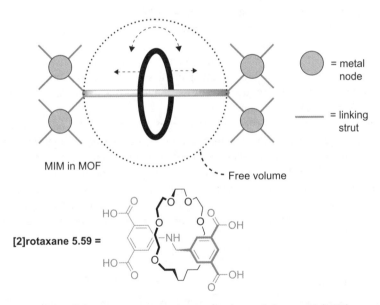

Figure 5.17 A [2]rotaxane is integrated into a metal–organic framework (MOF) where the free volume surrounding the macrocycle component ensures its dynamic properties remain intact (as indicated by arrows).

the door to switchable solid state devices that could miniaturize memory storage down to the nanoscale. We will now explore the fundamentals of dynamic MIMs in the next section.

5.4 Molecular machines

The dynamic properties of MIMs are perhaps their most widely exploited attributes. In essence, by restricting certain degrees of freedom, the mechanical bond provides a unique tool for controlling the relative positions of a MIM's constituent parts and directing their large amplitude molecular motion. This discovery has led to the development of a new field of research on artificial molecular machines, recognized by the award of the 2016 Nobel Prize in Chemistry. Molecular machines are molecular systems that perform tasks on the nanoscale using mechanical motion. Indeed, molecular machines are frequently employed in Nature; for example, myosin filaments fuelled by adenosine triphosphate (ATP) are operated in a ratchet-style mechanism to cause them to proceed over complementary actin filaments, to effect muscle contraction (Figure 5.18). In this section two categories of mechanically interlocked molecular machines, switches and motors, are defined and discussed separately. Particular attention has been paid to the blossoming applications of these machines as functional sensors, switchable catalysts, molecular pumps, and autonomous rotary motors.

The nature of the mechanical bond enforces unique steric constraints on the relative movements of MIM sub-components, defining a distinct set of large amplitude intramolecular motions. To envision these types of molecular motion, consider the degrees of freedom of a macrocycle on a static thread, where for a rotaxane or catenane the thread represents an axle or second macrocycle component respectively. The three dynamic processes available to the wheel are: i) its motion along the thread (translation); ii) its rotation about the thread (pirouetting); and iii) its swivelling on the thread (rocking); the first two being considerably less restricted and hence more useful (Figure 5.19). It should also be noted that in catenanes the translation motion of one ring is equivalent to the pirouetting of the second; therefore, by convention, it is the larger ring that is considered static (i.e. the thread), with translational motion frequently referred to as 'circumrotation'.

Analogous to the conformational isomerism arising from the restricted rotation of a single covalent bond, mechanical bonds generate stereoisomers known as co-conformations. A co-conformation defines the spatial arrangement of the MIM's covalent sub-units, with co-conformational interconversion occurring via the translational, pirouetting, or rocking motions of these components as constrained by the mechanical bond (Figure 5.19). At this point it is helpful to clarify that, alongside co-conformations, co-configurations (e.g. the enantiomers of [2]rotaxane **5.55** arising

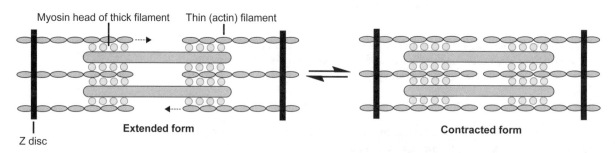

Figure 5.18 Schematic representation of muscle action in which contraction is actuated via a ratchet mechanism.

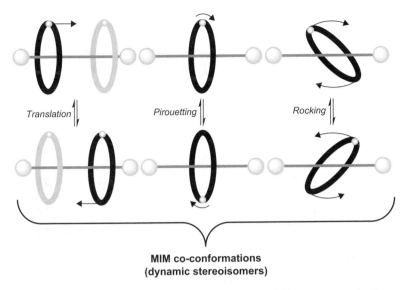

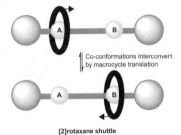

Translation   Pirouetting   Rocking

**MIM co-conformations
(dynamic stereoisomers)**

Figure 5.19 The three distinct types of molecular motion available to a macrocycle when constrained by a mechanical bond.

from mechanical planar chirality in Figure 5.14), are also stereoisomers exhibited by MIMs. On account of the molecular motion needed to switch between them, co-conformations are often considered dynamic stereoisomers, while co-configurations are static because they cannot usually interconvert unless a mechanical bond is broken. However, these distinctions can be blurred by dynamic rotaxanes that also generate mechanical planar chirality.

MIMs will often have non-covalent interactions between their sub-components that bias the distribution of dynamic stereoisomers, limiting them to a well-defined set of co-conformations. For example, in the cartoon of the bistable [2]rotaxane shown in Figure 5.20, the macrocyclic ring can reside at either recognition site A or B on the axle. Known as a molecular shuttle, this is a common blueprint design for MIMs exhibiting translational motion; thermal energy (i.e. Brownian motion) drives the translation of a ring (train) along an axle (track) between recognition sites (stations A and B), as constrained by the mechanical bond. Importantly, although molecular shuttles often refer to dynamic MIMs exhibiting translational motion, the concept is identical for co-conformations that interconvert by pirouetting.

Stoddart and co-workers reported the first molecular shuttle in 1991; a bistable [2]rotaxane (**5.60**) in which the 'blue box' π-electron poor macrocyclic wheel was shown to translocate between two identical π-electron rich hydroquinone axle stations at a rate of 1300 s^{-1} (Scheme 5.36). Since the axle component is symmetrical, the macrocycle has no thermodynamic preference for either station and so its position is equally distributed between them. Degenerate shuttles such as **5.60** provide fertile ground for probing the influence of molecular structure and environment on macrocycle shuttling kinetics. For example, the speed of macrocycle translocation in a bistable rotaxane can be reduced by: i) increasing the strength of non-covalent interaction(s) at stations; ii) introducing repulsive electrostatic interactions in the linker (i.e. speedbumps) or iii) increasing the

Co-conformations interconvert by macrocycle translation

[2]rotaxane shuttle

Figure 5.20 A bistable [2]rotaxane possesses two well-defined co-conformations that interconvert via the translational motion of the macrocycle component between two sites (stations A and B) on the axle component. MIMs of this type are known as molecular shuttles.

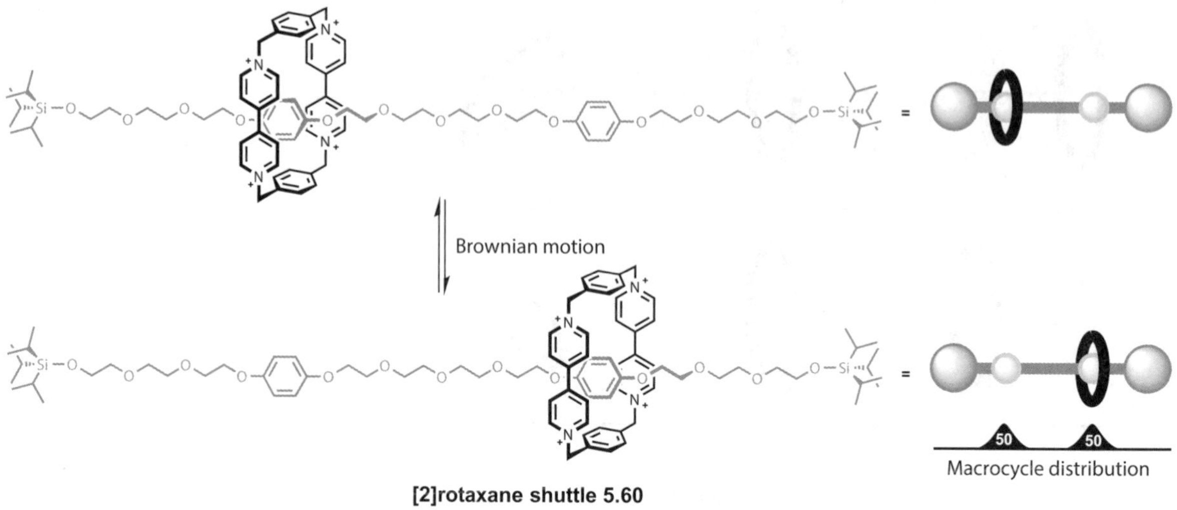

Brownian motion

[2]rotaxane shuttle 5.60

50   50

Macrocycle distribution

Scheme 5.36 The first molecular shuttle; a degenerate bistable [2]rotaxane in which the tetracationic macrocycle is equally located between two π-electron rich stations.

viscosity of the solvent medium. Interestingly, water has been shown to lubricate the translation and rotation of benzylic amide-based macrocycles (such as those in Scheme 5.13) by forming a stabilizing network of hydrogen bonds between the moving parts.

While in degenerate molecular shuttles (such as **5.60** in Scheme 5.36) the co-conformations are equally populated, a non-degenerate system biases the ring's position, creating a thermodynamic preference for one station over another (Scheme 5.37). Often this occupancy can be 'switched' using an external stimulus that modulates the non-covalent interactions between components. For example, in [2]rotaxane **5.61** the same electron poor macrocyclic wheel has a choice between two electron

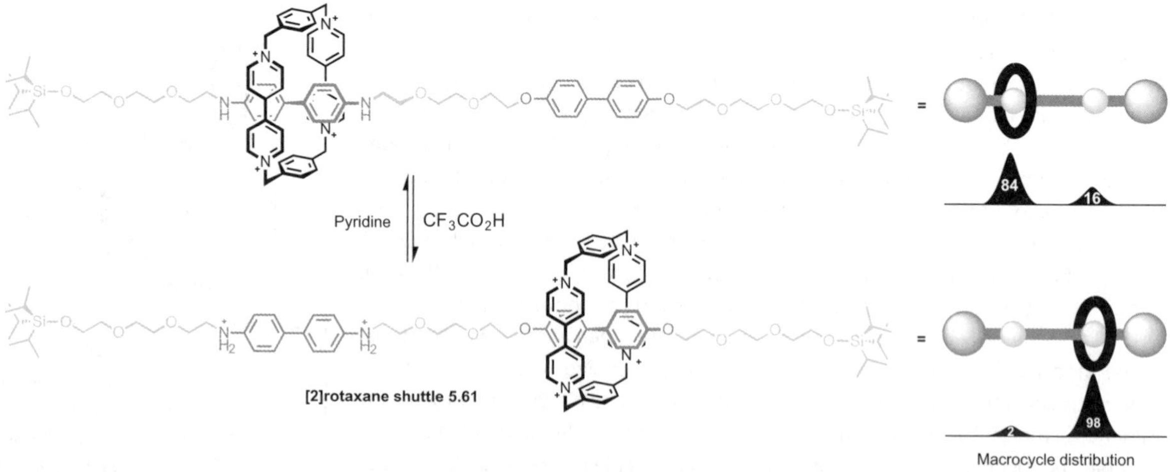

Pyridine ‖ CF$_3$CO$_2$H

84   16

[2]rotaxane shuttle 5.61

98

Macrocycle distribution

Scheme 5.37 A non-degenerate bistable [2]rotaxane can be actuated as a molecular switch; a pH stimulus is used to dictate the lowest energy co-conformation of the shuttle and thus the position of the macrocyclic wheel on the axle track.

rich stations, with the benzidine one being favoured at neutral pH due to stronger donor–acceptor aromatic stacking interactions (Scheme 5.37). However, protonation of this bis-amine creates a repulsive electrostatic interaction with the tetracationic macrocycle that switches its position, via translational motion, to favour the dioxybiphenyl recognition site, establishing a new lowest energy co-conformation. This concept forms the basis for molecular machines; a general description for dynamic, functional molecules that are key components in molecular nanotechnology. Coming under the umbrella of molecular machines are two broad subcategories of MIM devices; switches and motors, distinguished by the latter's ability to perform useful work. The subtleties of this distinction are explained below.

A molecular switch, such as the simple bistable rotaxane shuttle **5.61** (Scheme 5.37), cannot be used progressively to perform work since the change in position of the macrocycle on the axle influences the system only as a function of its state, in this case the pH. This means that the components of the shuttle are 'switched back' to their original position, as soon as the stimulus that drove them away from equilibrium is removed (i.e. benzidine protonation for **5.61**). Figure 5.21 outlines this concept using a simple thought experiment; a weightless train (macrocycle) and track (axle) are balanced on a see-saw, the position of which is dictated only by the weight (state) of stations (recognition sites) at either end. When a stimulus is applied to make B the heaviest station, the train rolls down the hill away from A. However, this weight imbalance is reversed when the stimulus is removed, causing the train to be returned to its starting point and undoing the progress of its journey (work). This represents the behaviour of a molecular shuttle.

By contrast, a **_molecular motor_** describes a system in which the train can engage a brake at station B so that it does not roll back down the hill upon see-saw inversion. Therefore, the train has moved away from station A irreversibly and, particularly for any passengers on-board, usefully! Returning to the molecular scale, this means that upon removing the stimulus, the macrocyclic ring will remain in an out-of-equilibrium

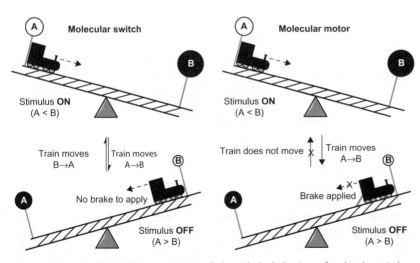

Figure 5.21 A simple thought experiment to distinguish the behaviour of molecular switches (left) and molecular motors (right), both types of MIM machines.

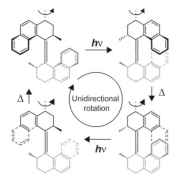

Molecular rotary motor 5.62

Figure 5.22 A combination of heat and light stimuli are used to generate unidirectional rotary motion in a non-interlocked molecular motor.

co-conformation, such that the energy put into the system has been used to perform work, in this case unidirectional molecular motion.

Finally, it is important to note that the term molecular machine does not only encompass mechanically interlocked molecules. Indeed, the 2016 Nobel Prize for the development of the 'world's tiniest machines', recognized Ben Feringa for his pioneering work on molecular rotary motors; molecules such as **5.62** that harness sterically restricted covalent bonds to direct their continuous rotation in the same direction (Figure 5.22). However, as pioneered by the other two recipients of this prize, Jean-Pierre Sauvage and Fraser Stoddart, dynamic MIMs are considered archetypal molecular machines, with imaginative new examples frequently reported in the literature. The following sections of this chapter outline exemplary MIMs from each category (switches and motors) as well as detailing their wide-ranging applications.

Mechanically interlocked molecular switches

The templation methods employed in synthesizing catenanes and rotaxanes almost inevitably result in interactions that live on between the interlocked components. In non-degenerate MIM shuttles, this generates recognition sites at which local intra-molecular interactions dictate co-conformational distributions, providing an exquisite tool for controlling dynamic behaviour. As such, systematic alterations to the recognition units, linkers, ring sizes, and topology of MIM shuttles have all been demonstrated to influence thermodynamically favoured co-conformations and the kinetics of their interconversion. Often the macrocycle's preferred station may be estimated by comparing association constants of model pseudorotaxane assemblies and then quantified using experimental techniques such as variable temperature NMR spectroscopy. For example, cooling down [2]rotaxane shuttle **5.61** (Scheme 5.37) reveals distinct sets of signals from the two translational isomers in its ^1H NMR spectrum (i.e. slow-exchange, see Section 1.5), the integration ratio of which is related to the proportion of rings occupying each station. It should be noted that while it is common to state that 'the macrocycle occupies station X', this terminology often means the position of equilibrium is biased towards that particular co-conformation.

A change in co-conformational population within a MIM switch occurs via large amplitude molecular motion, generating distinct states and precise switchable behaviour. As highlighted by the following examples in this section, switching can be actuated using a variety of physical (light, heat, pressure) and chemical (redox, pH, guest recognition) stimuli. Typically the stimulus turns off the non-covalent interactions stabilizing the lowest energy co-conformation, triggering population of the secondary binding site (i.e. the next lowest energy co-conformation), a perturbation to the system that is detectable using conventional analytical techniques. For switches, this process is reversed when the stimulus is removed.

While we have seen how pH can be used to dictate co-conformational change (Scheme 5.37), the competitive binding of a discrete chemical species, be it neutral or charged, (i.e. host-guest recognition) can also be exploited. The [2]rotaxane **5.63** is an anion-switchable molecular shuttle comprised of a pyridylbenzylic amide-crown ether macrocycle and a two-station axle component (Scheme 5.38). In the initial state the ring encircles the urea station because of strong inter-component hydrogen bonding interactions. As noted in Section 2.5, urea is also a potent anion

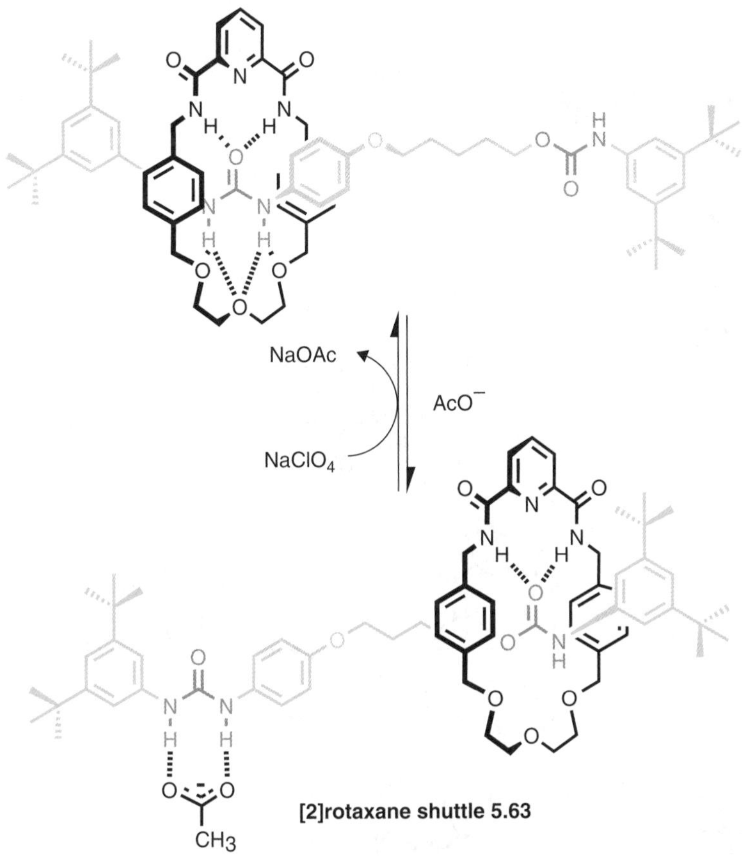

Scheme 5.38 A [2]rotaxane switch that is actuated through anion recognition.

[2]rotaxane shuttle 5.63

binding motif, hence the addition of acetate triggers macrocycle translation to the secondary carbamate station; a switch in co-conformation that can be characterized by ^{1}H NMR spectroscopy (protons are shifted upfield upon macrocycle encapsulation). Since anion binding is a reversible process, the system can be reset to its original co-conformation via the addition of $NaClO_4$ to precipitate NaOAc from the organic solvent.

In contrast to a non-degenerate molecular shuttle, **5.64** is a bi-stable [2]rotaxane containing two identical stations, each capable of binding a neutral guest (barbital) via complementary hydrogen bonding (Scheme 5.39). Molecular recognition at both stations forces the macrocycle to adopt the axle's central dodecamethylene chain as a 'station', essentially restricting the amplitude of its translational motion by destabilizing the starting point co-conformations. Degenerate MIM shuttles of similar design have revealed that molecular recognition can also affect the kinetics of macrocycle translation. For example, the chelation of Pd(II) between the axle and macrocycle components of [2]rotaxane **5.65** acts like a molecular brake, preventing the macrocycle from shuttling between the two triazole coordination sites, even upon heating (Figure 5.23).

Returning to the consideration of thermodynamics, it is important that there is a large energetic difference between co-conformations in each state, such that the

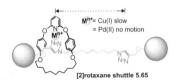

[2]rotaxane shuttle 5.65

Figure 5.23 In this [2]rotaxane shuttle the kinetics of macrocycle motion are altered by chelation to transition metal cations.

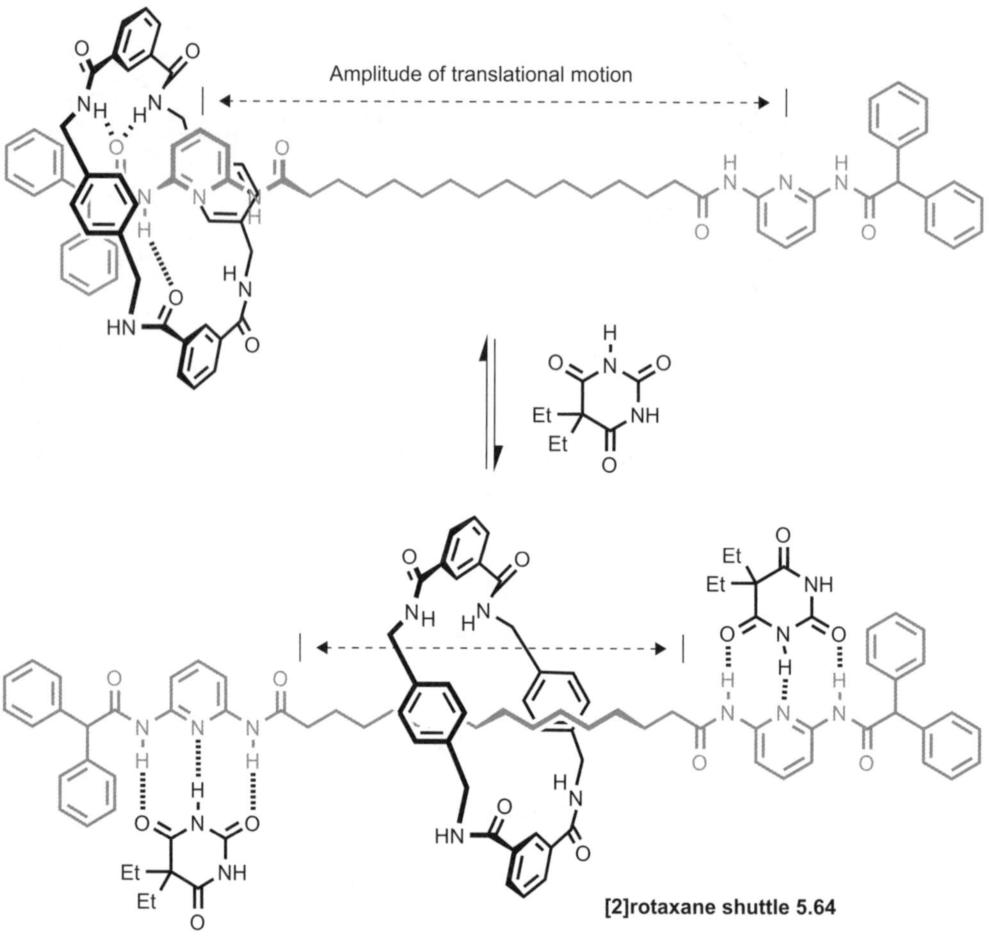

Scheme 5.39 The amplitude of macrocycle translation is restricted upon binding of barbital guests to the [2]rotaxane framework.

population of one is always heavily biased over the other and switching occurs with high positional integrity. In the previous molecular switches this was manifested in well-established binding sites for the macrocycle and strong guest recognition. A molecular switch with exceptional positional fidelity is bistable [2]rotaxane **5.66** in which an electrochemical stimulus is used to alternate between two states with co-conformational ratios of greater than 500:1 (Scheme 5.40). Initially superior hydrogen bonding with the axle succinamide station ensures its occupancy by the macrocycle. However, the position of equilibrium is overwhelmingly reversed when a one-electron reduction of the naphthalimide station substantially enhances its hydrogen bond accepting ability, triggering macrocycle relocation. Electrochemical oxidation restores the original binding affinities to reset the system. Incidentally, the MIM switch can also be operated photochemically with similar efficiencies, since excitation of the chromophoric station generates the naphthalimide radical anion in the presence of an electron donor. Indeed, light is also an effective stimulus to modulate the pirouetting motion of these benzylamide macrocycles. For example, photoinduced isomerization of the double

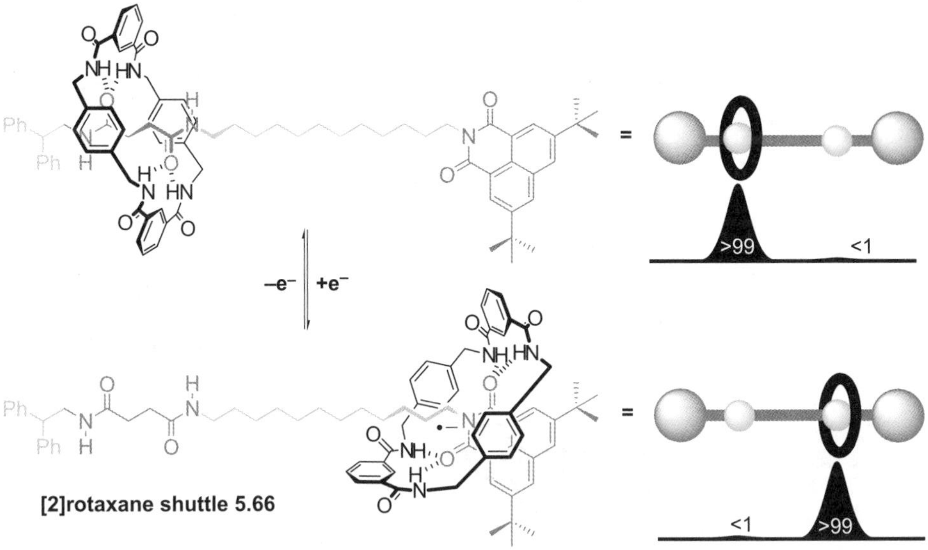

Scheme 5.40 An electrochemical [2]rotaxane switch with high positional fidelity.

bond in the axle component of [2]rotaxane **5.22** (Scheme 5.13) triggers motion by altering the strength of local intramolecular hydrogen bonding.

For molecular shuttles, electro- and photo-chemical stimuli are readily available, easily administered and do not normally generate chemical side products during switch operation. This is also true for MIMs that use solvent to induce a change in net position of the macrocycle. Solvent presents a powerful tool for co-conformational modulation because the intermolecular interactions governing MIM solvation are competitive with the non-covalent interactions between interlocked components. This phenomenon has been exploited to control the location of a pillar[5]arene macrocycle in [2]rotaxane shuttle **5.67** (Scheme 5.41). In chloroform the electron rich macrocycle encircles the polar imidazolium station, shielding it from the low polarity solvent. Increasing the ratio of DMSO, a more polar solvent, disrupts hydrogen bonding at this station and causes translation of the ring to the methylene groups neighbouring the carbamate stoppering unit, evidenced by upfield shifts of proton resonances in the ^{1}H NMR spectrum. In addition to their environment, MIM shuttles are often sensitive to temperature because, while the non-covalent interactions that anchor a macrocycle in one position are enthalpically favourable, an increase in ring mobility is favoured entropically. Therefore, it is important to consider the free energy change, i.e. $\Delta G = \Delta H - T\Delta S$, when rationalizing co-conformational switching in rotaxanes and catenanes. This relationship is the origin of entropy-driven shuttles; a large difference in entropy of macrocycle–station binding allows the temperature to have a significant impact on the population of a co-conformation. Indeed, heating [2]rotaxane shuttle **5.67** in DMSO switches the position of equilibrium back to the macrocycle occupancy at the imidazolium station (Scheme 5.40).

The translation and pirouetting of multiple interlocked components can be used in combination to produce exotic modes of molecular motion in higher-order MIM frameworks containing more than one mechanical bond. The actuation of these complex molecular shuttles has generated dynamic switching behaviour reminiscent

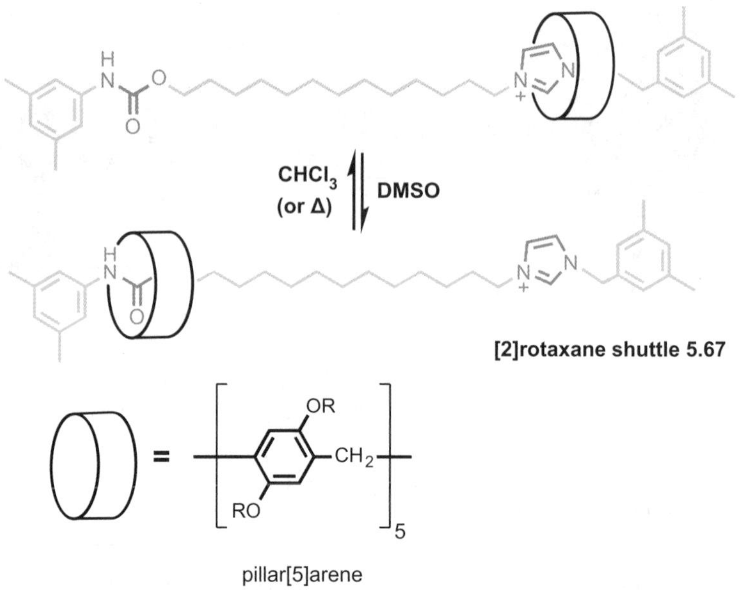

Scheme 5.41 A solvent- and temperature-directed [2]rotaxane molecular switch. R = solubilising groups.

of macroscopic machines such as muscles and pulleys. An example of the former is tristable rotaxane **5.68** in which mechanically bonded ring and axle components are also covalently bridged to give a daisy-chain architecture (Scheme 5.42). Starting from an elongated co-conformation, a variety of stimuli (pH, solvent, and temperature) can be used to gradually contract the rotaxane's geometry producing motion that is representative of muscle action. Meanwhile, the three mechanical bonds of triply interlocked [2]rotaxane **5.69** dictate bending of the axle's translational motion as it passes through the crown ether rings, producing linear and rotational motion that resembles a pulley (Figure 5.24). Further reports of molecular elevators, presses, cable cars, conveyor belts, and even the shuttling of a smaller ring through a larger one demonstrate that the scope of molecular motion, as facilitated by the mechanical bond, is limited only by one's imagination!

Applications of mechanically interlocked molecular switches

While many molecular switches act as prototypes towards molecular motors, the ability to switch between two well-defined co-conformations has yielded applications that include tuning guest recognition, promoting colorimetric sensing, and controlling chemical reactivity.

The bistable [3]rotaxane shuttle **5.70** acts as an adjustable receptor for bi-pyridyl guests (Scheme 5.43). Anchoring the two metallo-porphyrin-based macrocycles at the centre of the axle component favours the binding of shorter bidentate guests whilst switching to a state in which the rings are more labile enables the coordination of larger Lewis bases. Traditionally the binding of different sized guests would require a range of different sized synthetic hosts to have been prepared; however, [3]rotaxane **5.70** provides a single tuneable receptor that benefits from the induced fit (i.e. hand-in-glove) principle of binding.

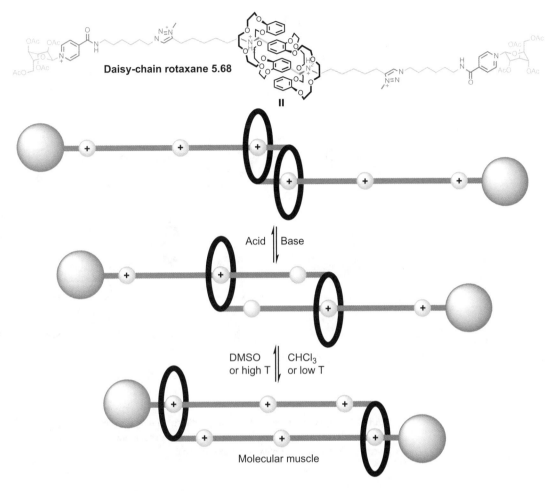

Scheme 5.42 The actuation of a daisy-chain rotaxane causes contraction and expansion of the MIM framework replicating the mechanical motion of muscle fibres.

Expanding the capabilities of rotaxane shuttles, the potential chirality of [2]rotaxane **5.71** has been exploited for enantioselective recognition (Scheme 5.44). When the central station of **5.71** is occupied the MIM is achiral; however, raising the pH triggers the asymmetric macrocycle to relocate to one of the two peripheral triazolium stations, generating two co-conformations that are mirror images of one another and so switching on mechanical planar chirality. In this state, a chiral (*S*)-camphor sulfonate anion guest binds preferentially to one enantiomer, biasing the position of equilibrium such that one chiral co-conformer is favoured over another (5:1 ratio). Distinct from a simple chiral shift reagent, where no biasing occurs, enantioselective recognition can be observed by ^{1}H NMR spectroscopy because the rotaxane shuttle forms diastereomeric host–guest complexes.

In addition to molecular recognition, the photoactive [3]catenane switch **5.72** in Scheme 5.45 affords anion sensing courtesy of a dynamic mechanism involving circumrotatory motion of its mechanically interlocked rings. The MIM elicits a colorimetric response for chloride because halide binding exposes the perylene diimide chromophore of the large central ring, disrupting its intramolecular aromatic stacking

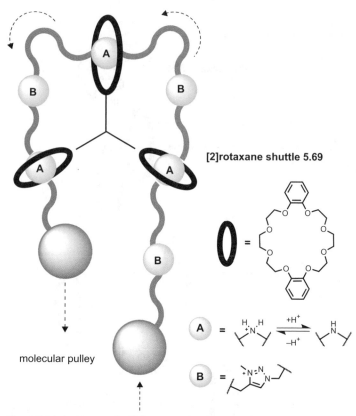

molecular pulley

Figure 5.24 A pH-controlled [2]rotaxane shuttle containing three mechanical bonds produces motion like that of a pulley when the chain moves to switch macrocycle occupancies between stations A and B.

interactions with the smaller peripheral macrocycles as they move to occupy the triazolium anion recognition stations. The amplitude of co-conformational change in switch **5.72** is reversible and large enough for it to induce a colour change visible to the naked eye; The fact that MIMs can display isomerism distinct from their non-interlocked counterparts has also been exploited for fluorescence emission switching. These shuttles may be employed in areas where the rapid switching of optical properties is required, such as devices for visible light communications.

Finally, [2]rotaxane **5.73** provides an elegant example of a molecular shuttle used as a switchable organocatalyst (Scheme 5.46). In this system the macrocycle component acts as a moveable steric shield, reversibly concealing and exposing a catalytic site on the axle component, namely a secondary amine that promotes the conjugate addition of a thiol to cinnamaldehyde. While the non-interlocked axle component is always active, the MIM catalyst can be switched on and off in response to an external stimulus (pH), mimicking the mechanism by which enzyme channels regulate substrate access to the active site. The MIM shuttle has since been upgraded to afford switchable enantioselective catalysis; the position of the macrocycle on the axle now dictates the 'handedness' of the chiral space around the active site and hence the enantiomeric product may be reversed without switching out the catalyst.

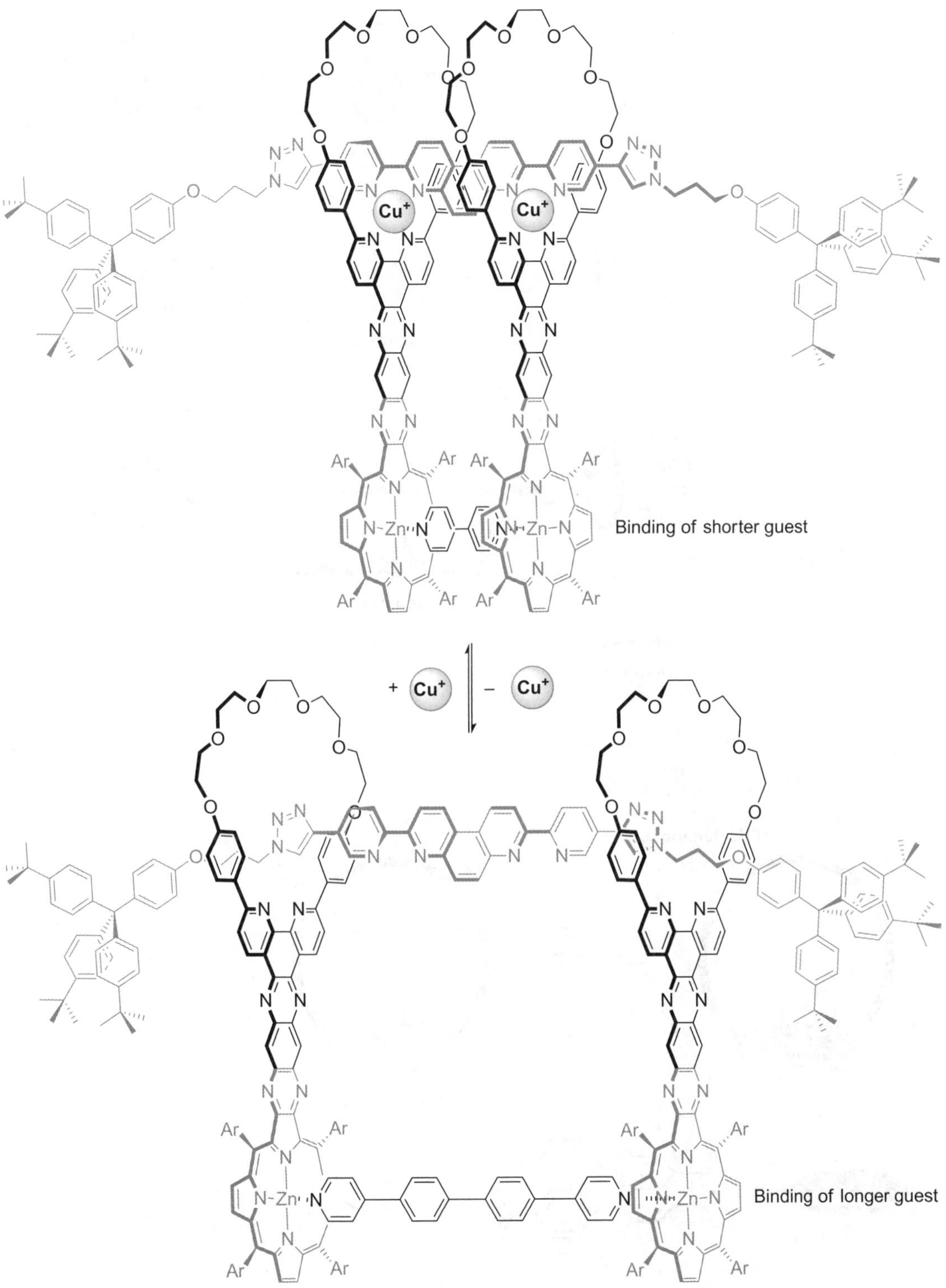

[3]Rotaxane 5.70

Scheme 5.43 The selectivity of this [3]rotaxane receptor can be tuned for longer or shorter bidentate guests by switching its dynamic properties on or off.

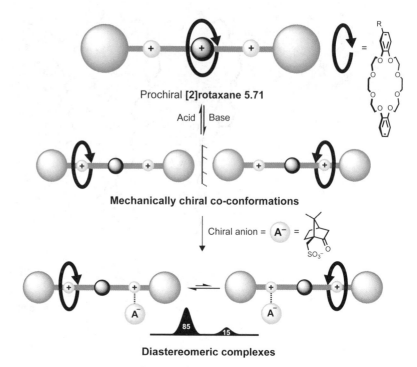

Prochiral **[2]rotaxane 5.71**

Acid ⇅ Base

Mechanically chiral co-conformations

Chiral anion = **A⁻** =

Diastereomeric complexes

Scheme 5.44 Switching on the dynamic behaviour of a [2]rotaxane generates two co-conformations that are also chiral co-configurations due to mechanical planar chirality. This enables the enantioselective recognition of a chiral anion.

[3]catenane 5.72

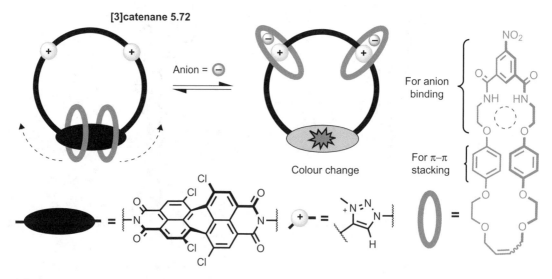

Anion = ⊖

For anion binding

For π–π stacking

Colour change

Scheme 5.45 The anion recognition-induced circumrotatory motion of a [3]catenane is concomitant with a colour change, enabling it to act as an optical anion sensor.

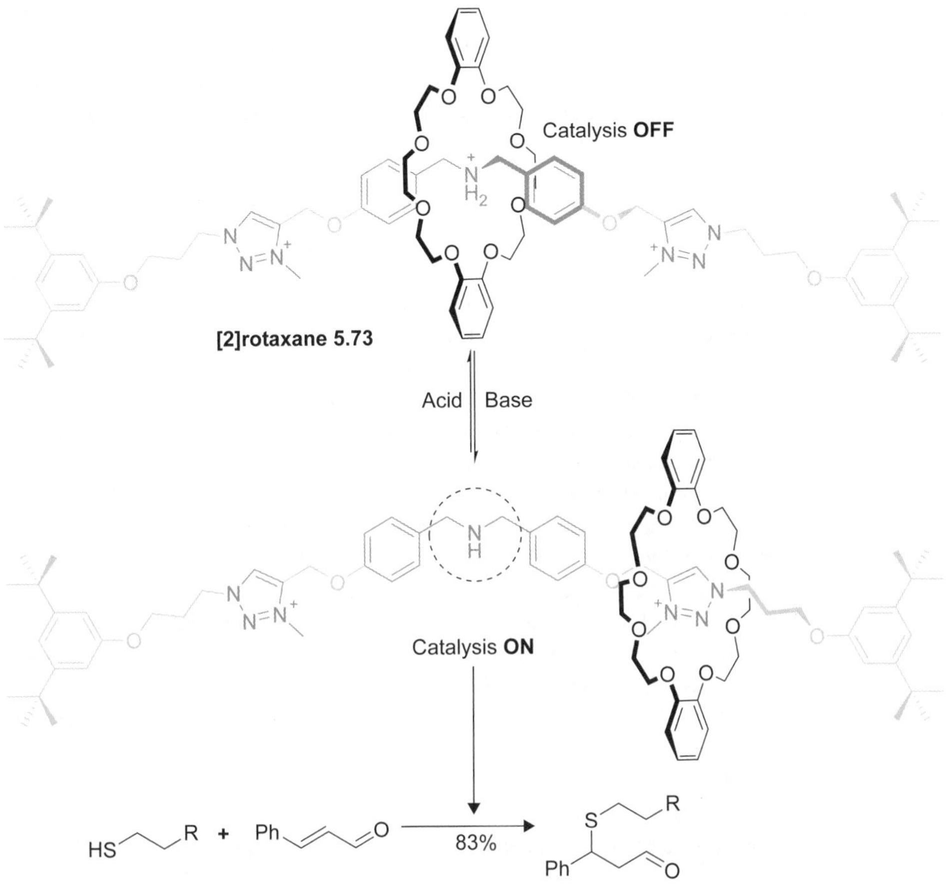

Catalysis OFF

[2]rotaxane 5.73

Acid | Base

Catalysis ON

83%

Scheme 5.46 Switching the macrocycle's position in a [2]rotaxane (pH stimulus) enables its catalytic behaviour to be turned on and off.

In addition to the many uses of switches as discrete molecules in solution, their assembly into ordered arrays holds much promise for the fabrication of MIM-based responsive materials and devices. For example, rotaxane shuttles have been appended to silica nanoparticles; these act as nanovalves dictating the uptake and release of a chemical cargo such as drug molecules, affording targeted delivery to reduce the dosage required.

Mechanically interlocked molecular motors

In the MIM switches described so far, machine operation occurs under thermodynamic control meaning that the mechanical energy used to drive the macrocycle away from equilibrium is lost upon removal of the stimulus, as the ring moves back to its original position. To function as a molecular motor, the biased Brownian motion that results from driving the machine away from equilibrium must be harnessed for work before the system re-equilibrates. This is how biological molecular machines operate. For example, we saw earlier how the hydrolysis of ATP fuels myosin filaments to pass over complementary actin filaments to effect muscle contraction (Figure 5.18). Importantly, in our bodies muscle relaxation is actuated on-demand and not simply on removal of the stimulus (ATP), as is the case for the molecular muscle-like switch **5.68** in Scheme 5.42.

A key challenge to the design and operation of synthetic molecular motors is one of scale; essentially all molecular-scale motion must constantly compete with background 'thermal noise'. Returning to the earlier analogy of a macroscopic machine (Figure 5.21); whilst a moving train is unaffected by the random Brownian motion of molecular-sized particles in its path, place it in a hurricane, where now the wind's energy is competitive with that of the train, and the 'background noise' presents a significant barrier to the train's progression in a straight line. This describes the situation for molecular machines and, in part, explains why artificial molecular motors are fewer in number than switches. However, this is a fertile area in modern research in which the development of out-of-equilibrium machines can deliver functional systems that can perform meaningful tasks. The extraction of work from nanoscale motion is conceptually challenging and so this section begins by outlining key concepts using a relatively simplistic molecular motor before highlighting two examples that realize ultimate molecular machine behaviour.

In a switchable molecular shuttle (Section 5.4) the macrocycle can always seek its equilibrium position on the axle; however, molecular motors often employ kinetic barriers to prevent this backward motion. This compartmentalizes the macrocyclic ring on the axle track, such that kinetic control (i.e. the rate of exchange between stations) is combined with thermodynamics (i.e. the modulation of relative binding affinities) to produce a more sophisticated machine that can perform useful work. The [2]rotaxane **5.74** in Scheme 5.47 is one such system. This MIM is able to change the position of its macrocycle irreversibly, operating via a fundamentally different mechanism to a molecular switch. The starting co-conformation has the ring residing at the more favourable fumaramide station (state A); however, photoisomerization of the alkene destabilizes the macrocycle's position, making the succinamide site the preferred thermodynamic

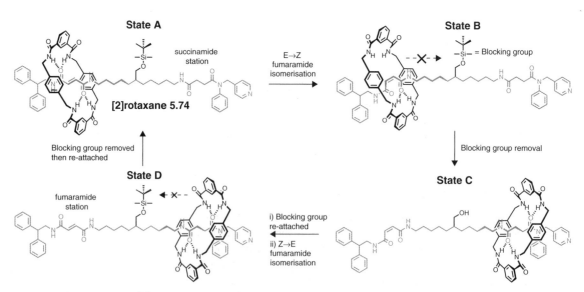

Scheme 5.47 An irreversible [2]rotaxane ratchet where a bulky silyl ether group is used to kinetically trap the macrocycle in a thermodynamically unfavourable position (states B and D) allowing unidirectional motion (on transition to states C and A respectively).

location (state B). This creates a potential energy gradient for directional motion of the macrocycle that, in this case, is only actuated upon removal of the bulky silyl ether group. Cleavage of this kinetic barrier allows macrocycle translation; biased Brownian motion establishes the new equilibrium position in which the ring now occupies the succinamide station (state C). The system is then reset by replacing the barrier and re-isomerizing the alkene (state D). However, this does not undo the change in macrocycle positon until the barrier is removed for a second time, allowing it to return to the initial location (A). Notably, in states B and D the macrocycle has been moved energetically uphill but remains kinetically trapped, a mechanism known as ratcheting that is characteristic of many biological machines, including muscle fibres (Figure 5.18). This enables [2]rotaxane **5.74** to produce the directional and repeatable transport of a macrocycle, key characteristics of a progressive molecular motor.

Further exploration into ratcheting MIMs has seen the development of molecular machines that can pump macrocycles energetically uphill by capturing them on an axle component, forming an [*n*]rotaxane (**5.76**, Scheme 5.48). The axle **5.75** has a dumbbell structure, except that the dimethylpyridinium group on one end is small enough to allow slippage of the tetracationic 'blue box' macrocycle **3.22** onto **5.75** upon electrochemical reduction, driven by non-covalent association of the resulting radical cations (state A). This generates a pseudorotaxane assembly which, upon re-oxidation,

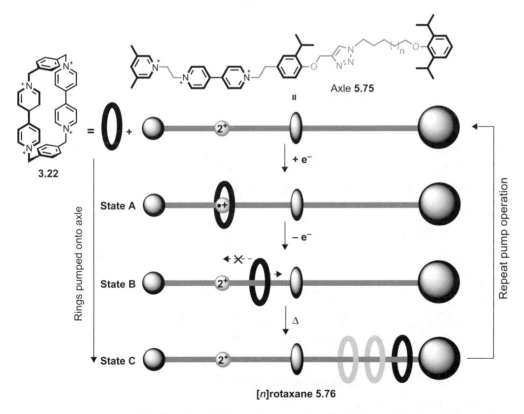

Scheme 5.48 A system in which electrochemical energy is converted to chemical potential energy by the pumping of macrocycles uphill where they are stored using an [*n*]rotaxane architecture.

introduces an electrostatic barrier in the form of the bipyridinium unit that prevents ring de-threading (state B). The axle component also contains a bulky isopropylphenol group that, now neighbouring the macrocycle, presents a steric barrier to its movement into the next compartment of **5.75**. By design, this steric barrier is lower in energy than the electrostatic one, so that heating the system in state B enables directional translation of the ring over this hurdle and onto the oligomethylene chain where, at room temperature, it becomes kinetically trapped, forming [2]rotaxane **5.76** (state C). These steps can be repeated to enable further rings to be loaded onto the axle in high efficiency. Importantly, there are no significant stabilizing intramolecular interactions in this compartment, meaning rings have been driven uphill in energy to a higher local concentration on the axle in the [n]rotaxane. The mechanical bonds are responsible for the storage of these macrocycles. Therefore, this molecular machine is a pump, performing work in the manner of a carrier protein that actively transports substrates against a concentration gradient.

The operation of both previous molecular machines required a sequence of chemical, electrochemical, or thermal stimuli to be applied, in a correct order, to perform their mechanical work task. However, biological machines operate autonomously, making use of a continuous supply of chemical energy, typically ATP, to fuel their moving parts. Taking inspiration from this, [2]catenane molecular motor **5.77** is capable of continuous, uni-directional motion of a smaller ring around its larger one via the irreversible consumption of a chemical fuel (Scheme 5.49). As in [2]rotaxane **5.74** (Scheme 5.47), the track contains recognition sites and removable blocking groups to create transient compartments for the smaller mobile ring. A key difference to **5.74** is that the choice of blocking group for **5.77**, 9-fluorenylmethoxycarbonyl (Fmoc), ensures that the repeated raising and lowering of kinetic barriers, as required for directional (circumrotatory) motion, can occur under a single set of reaction conditions. Importantly, the different mechanisms responsible for Fmoc barrier cleavage and re-attachment

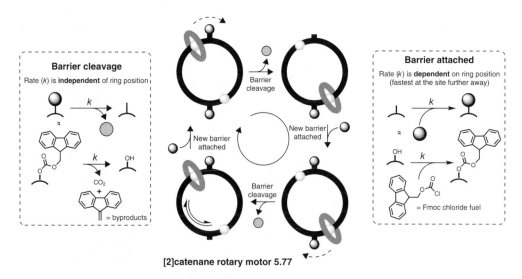

[2]catenane rotary motor 5.77

Scheme 5.49 A state-of-the-art MIM motor, this [2]catenane is capable of unidirectional and autonomous rotary motion upon consumption of a chemical fuel.

cause the latter process to be faster when the macrocycle is far from the reactive site, ensuring the ring will always enter the next compartment from the same direction. This affords net directional rotation of the molecular motor for as long as unreacted fuel (the electrophile Fmoc chloride) remains present in solution. Mimicking biological machines that are also powered by continuous chemical reactivity, [2]catenane **5.77** acts as truly autonomous molecular motor that may be employed as an engine to perform tasks on the atomic scale, potentially biological transport, chemical synthesis or data storage.

5.5 Summary and conclusions

In a class of their own, mechanically interlocked molecules are formed when two or more molecules become entangled within one another and cannot come apart due to the inability of bonds to pass through one another. Importantly, the molecular sub-components of MIMs are not covalently bonded together but are instead connected via a mechanical bond. While mechanical bonds are common in the macroscopic world and perform a number of roles in important biological processes, artificial MIMs are relatively new, their arrival attributed to Sauvage's landmark discovery of the metal template-directed synthesis of a [2]catenane (Scheme 5.3). Now armed with modern-day synthetic techniques and the guiding principles of supramolecular chemistry, a plethora of catenane topologies and rotaxane architectures have been developed, often containing multiple mechanical bonds and rivalling the supramolecular assemblies from Chapter 4 in scale.

From the first reports of synthetic MIMs it was evident that the mechanical bond has a strong influence on chemical and physical properties. Restrictions to the relative motions of molecular components, increased steric hindrance around embedded functional groups and alterations to the non-covalent interactions between constituent parts are all common outcomes of mechanical bonding. As such, the MIMs outlined in this chapter are finding applications across multiple scientific disciplines, including as selective receptors, sensory devices, chiral catalysts, conductive wires, optical probes, and flexible materials.

The unique dynamic properties of MIMs have been critical to their enormous impact within the field of artificial molecular machines, as both switches and motors. While the first [2]rotaxane molecular shuttle was only reported at the end of the last century, a wealth of rotaxane- and catenane-based machines, actuated using a wide variety of chemical, light, and redox stimuli, have since been developed.

A pinnacle of artificial molecular machine behaviour is David Leigh's peptide-synthesizing [2]rotaxane **5.78** in Scheme 5.50, a molecule that can make other molecules. During the operation of **5.78** a macrocycle bearing an active organocatalyst (cysteine residue) moves in one direction (from left to right) along its molecular thread cleaving three amino acid units and transferring them to a reactive terminus (amino group). The output from this process is a single, sequence-specific polypeptide and hence this rotaxane replicates the complex functionality of an archetypal biological machine, the protein-synthesizing ribosome.

While the desire for chemists to prepare mechanically interlocked molecules may have begun as no more than chemical curiosity, rotaxanes and catenanes have, quite literally, set the wheels in motion for an industrial revolution on the nanoscale!

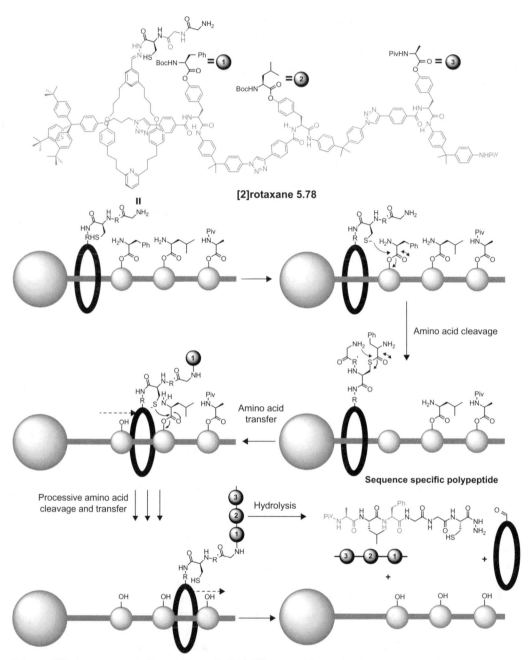

[2]rotaxane 5.78

Amino acid cleavage

Amino acid transfer

Sequence specific polypeptide

Processive amino acid cleavage and transfer

Hydrolysis

Scheme 5.50 A sequence-specific peptide-synthesizing [2]rotaxane. The catalyst-bearing macrocycle cleaves each amino acid blocking group in turn as it proceeds along the axle, transferring it to the *N*-terminus and so building up the polypeptide chain.

5.6 Further reading

The following resources provide further examples and applications of mechanically interlocked molecules.

- An exhaustive collection of molecules containing mechanical bonds, including their synthesis, properties and applications is provided in the book: Bruns, C. J. and Stoddart, J. F., 2016, *The Nature of the Mechanical Bond*, New York: John Wiley & Sons, Inc. ISBN: 978-1-119-04400-0.

- The following comprehensive reviews detail the history of artificial MIMs, including; i) catenanes *Angew. Chem. Int. Ed.* 2015, **54**, 6110–50, ii) rotaxanes *Chem. Rev.*, 2015, **115**, 7398–501, iii) molecular knots *Angew. Chem. Int. Ed.*, 2017, **56**, 11166–94, and iv) topologically complex molecules in general *Chem. Rev.*, 2011, **111**, 5434–64.

- The influence of the mechanical bond on the properties of MIMs and their emerging applications as functional molecules and materials are described in the following review articles: *Chem. Commun.*, 2014, **50**, 5128–42, *Chem. Commun.*, 2017, **53**, 298–312, and *Chem. Soc. Rev.*, 2019, **48**, 5016–32.

- The dynamic properties of MIMs, including their roles as artificial molecular machines, are detailed in publications: *Angew. Chem. Int. Ed.*, 2006, **46**, 72–191, *Chem. Rev.*, 2015, **115**, 10081–206, and *ACS Cent. Sci.*, 2020, **6**, 117–28. Video recordings of the Nobel Lectures on molecular machines are available via the following weblinks: J. P. Sauvage (https://www.nobelprize.org/prizes/chemistry/2016/sauvage/lecture/), J. F. Stoddart (https://www.nobelprize.org/prizes/chemistry/2016/stoddart/lecture/), and B. L. Feringa (https://www.nobelprize.org/prizes/chemistry/2016/feringa/lecture/)

5.7 Exercises

5.1 (a) What are the differences and similarities between rotaxanes and catenanes in terms of mechanical bonding and chemical topology?

(b) What are the implications of these differences in terms of the strategies used to make catenanes and rotaxanes? In cartoon format, sketch out some examples of the most common approaches.

(c) Outline which interactions have been harnessed for the template-directed syntheses of MIMs and under what conditions they are most appropriate. Do strong intermolecular interactions with a template automatically deliver a high yield of the MIM?

(d) What are the advantages of using active metal templates over passive metal templates?

(e) Besides Cu(I) azide-alkyne catalysed cycloaddition chemistry, suggest some other metal-catalysed chemical transformations that might be appropriate for active metal-templated synthesis of a [2]rotaxane and outline the key requirement for this synthetic strategy.

5.2 (a) In cartoon format, name and sketch the three types of motion available to a macrocyclic ring on a molecular thread. How are the two most common types of motion related to one another in a [2]catenane?

(b) The potential energy profile of the bistable [2]rotaxane below indicates it is a degenerate molecular shuttle. Sketch a potential energy profile for the scenario in which this [2]rotaxane is a non-degenerate molecular shuttle, favouring station A.

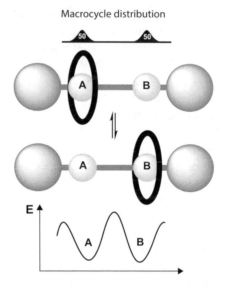

(c) In the non-degenerate scenario, the co-conformational preference of the [2]rotaxane shuttle is switched for station B by the cooperative binding of a guest between macrocycle and axle components. How is this change reflected in your potential energy diagram? What can be said about the possible influence of guest binding on the kinetics of the [2]rotaxane shuttle (at constant temperature)?

(d) What other stimuli may be exploited to induce co-conformation change in [2]rotaxane shuttles? How might the change in co-conformation be: i) character-ized, and ii) quantified?

(e) How could the [2]rotaxane switch be converted into a molecular motor?

For the answers to these exercises, visit the online resources which accompany this primer.

Glossary

Chapter 1

Binding affinity A measure of the strength of interactions between a host molecule and its guest.

Chelate effect The effect where binding affinity of a host towards a guest is enhanced by the presence of multiple binding sites on the host, driven by enthalpic and entropic factors.

Complementarity The relationship between two interacting molecules where both molecules have the opposite donor/acceptor properties (e.g. hydrogen bond donor-acceptor, opposite electrostatic charges) such that they are able to attract each other without generating internal molecular strains.

Cooperativity The phenomenon where guest binding on a site on a receptor influences the affinity of other sites on the same receptor towards guest species.

Cryptate effect An extension of the chelate and macrocyclic effects where guest affinity is further enhanced in a 3D ligand possessing multiple binding sites.

Hydrophobic effect The tendency for non-polar and hydrophobic molecules (typically unable to form hydrogen bonds), functional groups, and molecular surfaces to preferentially aggregate in water, in order to reduce their unfavourable interactions with water molecules.

Macrocyclic effect The higher affinity of cyclic molecules towards guests compared to their acyclic analogues.

Preorganization The phenomenon where the binding site of a host molecule is held in a conformation that already pre-disposes it towards interaction with the guest, without the need to undergo further structural deformation in order to conform to the guest's 3D shape.

Solvophobic forces Factors accounting for the poor miscibility or solubility of compounds possessing drastically different polarities (e.g. non-polar vs polar solvents), due to the inability of the solute–solvent interactions to overcome the strength of solvent–solvent interactions.

Chapter 2

Allosteric binding The phenomenon where binding of one species changes the conformation of the receptor structure that affects the binding affinity at a separate site towards a different guest species.

Calix[n]pyrroles Macrocycles which comprise of n pyrrole units which are highly effective in anion binding primarily by hydrogen bonding to the pyrrole N–H donors.

Cryptands A family of synthetic bicyclic and polycyclic multidentate ligands for binding cations. Their affinities exceed those of analogous macrocycles due to the *cryptate effect*.

Heteroditopic A receptor possessing binding sites for both cations and anions.

Hofmeister series A classification of ions originally based on their abilities to increase or decrease the aqueous solubilities of proteins.

Lariat ethers Crown ethers with pendant arms containing additional Lewis basic groups to further strengthen the coordination of cations.

Phase transfer catalysts Molecules present in small sub-stoichiometric (i.e. catalytic) quantities which facilitate the physical migration of molecules or ions across the boundaries between two immiscible phases, usually an organic and an aqueous phase.

Siderophores Naturally occurring molecular receptors secreted by microorganisms such as bacteria and fungi which can bind iron(III) cations very strongly, facilitating their transport across cell membranes.

Zwitterion A molecule which contains an equal number of positively and negativelycharged groups.

Chapter 3

Carceplex The complex formed when guest molecules are imprisoned within the hollow concave cavities of molecular cages called carcerands.

Chiral discrimination The process where a chiral molecule binds different guest enantiomers or diastereoisomers with different affinities, thus discriminating between them.

Covalent organic framework A class of 2D or 3D solid porous materials comprising of building blocks linked together by strong covalent bonds.

Differential sensing a strategy of molecular sensing where an array of receptors is simultaneously exposed to the guest molecule, which generates a unique sensing pattern or identification signature for each guest molecule based on each individual receptor's affinity for the guest.

Hemicarcerands Molecular containers with openings large enough to allow entry and exit of one or more guest molecules from their concave internal cavities at elevated temperatures. These differ from carcerands which permanently entrap the guest molecule(s) within their cavities which cannot be released without the destruction of the carcerand.

Hydrogel A network of crosslinked or self-assembled molecules or polymers which are able to entrap a large quantity of water to form a solid structure.

Inclusion complexes A supramolecular complex where a host possessing a hollow cavity encapsulates another molecule (the guest) within it.

Metal–organic framework: a class of crystalline and often porous material that consists of coordination bonds between transition metal cations and multidentate organic linkers.

Chapter 4

Amphiphile A molecule that contains distinct polar (hydrophilic) and non-polar (hydrophobic) regions, often leading to their self-assembly into ordered structures (such as bilayers, micelles, and vesicles) in solution.

Foldamer An oligomeric molecule that folds into an ordered, compact conformation in solution.

Helicate A helical supramolecular complex, typically formed by the self-assembly of multidentate organic ligands that are wrapped around ions positioned along the helical axis.

Self-assembly The spontaneous and reversible association of molecules and/or ions to form a larger, specific, and more complex supramolecular structure.

Supramolecular polymer These are polymeric materials formed from monomers held together by non-covalent interactions and hence they maintain their polymeric properties in solution.

Thermodynamic product This is the product with the lowest free energy on a reaction energy profile, typically favoured under reversible reaction conditions.

Chapter 5

Catenane A molecule that contains two or more mechanically interlocked macrocyclic rings. Catenanes have a non-trivial *chemical topology*.

Chemical topology Refers to molecules that contain crossing points when their structures are represented in a plane. Molecules are said to have non-trivial chemical topologies if crossing points are not broken upon continuous deformation e.g. a catenane.

Co-conformation Refers to the spatial arrangement of molecular components in a mechanically interlocked molecule. Interconversion between co-conformations occurs via translation, pirouetting, or rocking molecular motion.

Mechanical bond An entanglement between two or more molecular components that cannot be separated without breaking or

distorting chemical bonds. *Catenanes* and *rotaxanes* both contain one or more mechanical bond.

Mechanically interlocked molecule Molecules such as rotaxanes and catenanes that contain entangled molecular components. These components cannot be separated without breaking or distorting chemical bonds.

Molecular machine A general description for a dynamic, functional, and, in this book, mechanically interlocked molecule that, upon exposure to a stimulus, undergoes large amplitude mechanical motion of one molecular component relative to another.

Molecular motor A type of molecular machine that can perform work because removal of the stimulus does not return the molecular components to their original position. An example is a catenane that exhibits unidirectional rotary motion of one macrocyclic ring around another.

Molecular switch A type of molecular machine in which molecular components are returned to their original position as soon as the stimulus that switched them is removed. An example is a rotaxane shuttle in which the macrocyclic ring can always occupy its equilibrium position on the axle.

Pseudorotaxane An interpenetrated assembly formed by non-covalent interactions between a molecular ring and thread components. The resulting supramolecular complex resembles a rotaxane without the stoppering groups.

Rotaxane A molecule comprising at least one macrocycle that is threaded by a linear component (axle) terminated by bulky end groups (stoppers) to prevent ring dethreading. Relative to catenanes, rotaxanes have a trivial *chemical topology*.

Index